Homo dominus
A Theory of Human Evolution

Stephen G. Dennis

iUniverse, Inc.
New York   Bloomington

Homo dominus
A Theory of Human Evolution

iUniverse books may be ordered through booksellers or by contacting:

iUniverse
1663 Liberty Drive
Bloomington, IN 47403
www.iuniverse.com
1-800-Authors (1-800-288-4677)

ISBN: 978-0-595-53125-7 (soft)
ISBN: 978-0-595-63187-2 (ebk)

Library of Congress Control Number: 2008943688

Printed in the United States of America

iUniverse rev. date: 1/12/2009

Dedication

For Sherry, who makes all things possible,

for Christopher, who makes them fun,

and for my family and teachers, who have waited so patiently.

The controlling intelligence understands its own nature,

and what it does, and whereon it works.

— Marcus Aurelius, Meditations, VI.5

Table of Contents

Author's Note   xi

Preface   xiii

Homo dominus, the Controller: The Human Strategy   1
   Prologue   1
   Framing the Question   4
   Homo dominus Overview   5
   The Nature of Control   6
   Control and the Quest for Stability   13
   The Fundamental Divergence   16
   Control, Stability, and Human Evolution   17
   The Human Signature   23
   Homo dominus and the Human Strategy   25

Homo auguris, the Seer: Thought in a Changing World   27
   I Think, Therefore I Survive   27
   Fundamentals of Cognition   28
   Neurophilosophical Linkages   33
   Hierarchies and Their Problems   35
   Cognition and Control   41
   The Emergence of the Seers   45
   The Unique Human Brain   51
   Homo auguris and the Human Strategy   54

Homo ipsianimus, the Self-Aware: A Circle Game   56
   What Cognition Feels Like   56
   Introspection   56
   The Circular Logic of Self-Awareness   60
   The Circular Logic of Control   64
   Looking for Loops   65
   The Equivalence Principle   66
   On Emotions   67
   Homo ipsianimus and the Human Strategy   70

Homo conlocutus, the Converser: Sharing Thoughts   72
   Are You Talkin' to Me?   72
   The Adaptive Value of Language   73
   Linguistics and Universal Grammar   75
   Protolanguage   78
   Nouns and Verbs, Perceptions and Actions   81
   Nouns and Verbs in Brains   84
   The Language of Control   89
   Language and Social Organizations   95
   Homo conlocutus and the Human Strategy   96

Homo habilis, the Technologist: Hands, Fingers, Knees, Toes   98
   Brain or Brawn?   98
   Bipedal Locomotion   99
   Secondary Altriciality   102
   The Hominid Trinity   105
   The Transportation Economy   109
   Tools, Technologies, and Control Loops   110
   The Hungry Brain   115
   The Technology of Meat   117
   Homo habilis and the Human Strategy   119

Homo bellicosus, the War-Maker: A New Recipe for Extinction   121
   The Dark Side   121
   The Structure of Social Groups   123
   Sexual Interactions   125
   Kinship Interactions   128
   Inlawship and Friendship Interactions   131
   Combining and Optimizing: SKIF Communities   132
   Xenophobic Interactions   135
   Human Aggression and Control   138
   Atrocity and Genocide   141
   Homo bellicosus and the Human Strategy   143

Homo beneficus, the Altruist: The Kindness of Strangers   145
   Bang the Drum Slowly   145
   A Perfect World   147
   Sexual/Reproductive Altruism   150
   Kinship Altruism   151
   Inlaw Altruism   152
   Friendship Altruism and Reciprocity   154
   Xenophobia and Group Selection   157

Cultural Altruism   159
The Control Interpretation of Altruism   160
Homo beneficus and the Human Strategy   163

Homo humanitas, the Enculturated: A Sea of Trivialities   165
Taming the SKIFs   165
Nature-Nurture Redux   166
Memes and Genes   169
Trivialities   172
Cultural Signaling and the Control of Xenophobia   172
The Control Interpretation of Cultural Signaling   180
Lies, Deceptions, and Viral Memes   183
Homo humanitas and the Human Strategy   185

Homo aestheticus, the Artist: The Art of Losing Control   187
The Problem of Art   187
Beyond Memetics   188
The Emergence of Artists   189
The Appreciation of Art   191
Mechanistic Interpretations   193
Homo aestheticus and the Human Strategy   197

Homo mortalis, the Mortal: Controlling the Uncontrollable   199
Stepping Between, Never On, the Lines   199
Uncertainty   200
Death   202
Spirituality   205
Religion, Science, and Sibling Rivalry   210
The First Inklings   212
Homo mortalis and the Human Strategy   214

Homo sapiens, the Wise: *Quo Vadis?*   216
The Path to Homo sapiens   216
The Paleontology of Control   217
The Future of Control   221
The Case for Altruism   222
The Case for Culture   223
The Case for Religion   226
The Case for Science   228
Accepting Ourselves   231

Author's Note

This book marks the third phase in my rather unconventional career. The first was in scientific research, starting in the late 1960s at MIT, where I earned a Bachelor's degree in biology. After MIT, I did graduate work in behavioral genetics with Seymour Benzer at Caltech, and then in physiological psychology with Tony Deutsch at the University of California/San Diego, where I earned a doctorate. I moved to Montreal in 1975 to do postdoctoral research in pain physiology and opiate analgesia with Ron Melzack at McGill University, then to Boston to take a staff scientist position at the Neurosciences Research Program, an MIT research center. There, I had the opportunity to work with some of the best minds in neuroscience, co-authoring a monograph with Jean-Pierre Changeux on signal transduction across biological membranes, and co-editing a book on the structure of the cerebral cortex. In 1980, I moved to Seattle to take a position as Research Assistant Professor at the University of Washington School of Medicine. Unfortunately, funding cutbacks in basic research put me in the difficult position of having to choose between the career I wanted and the place I loved.

In the end, I decided to stay in Seattle and postpone my scientific career. Phase two began when I took a management position at USWEST (now Qwest), one of the baby Bells formed from the breakup of AT&T. I quickly established myself in strategic planning and new product development, starting in the cellular division, and then moving to public phones, advanced wireless technologies, video on demand, smart cards, and e-commerce. In a short time, twenty years passed, and I had built a successful and distinguished career.

Now, you might think two such disparate human activities as telecommunications and neuroscience would have little or no connection, but you would be mistaken. What links them is the root concept of *Homo dominus*, something I realized, bit by bit, during the long daily commutes to and from downtown Seattle. Hopefully, it is something you will realize as you read the following pages. As the implications of this concept became clear, I returned to my scientific roots and once again immersed myself in the literature. It soon became evident that the scientific foundations of *Homo dominus* were already in place, just not called out as such. This book seemed the logical next step, and thus began phase three.

I wrote *Homo dominus* primarily for first-year graduate students in the neurosciences, social sciences, biology, and philosophy, plus various other connected disciplines. Of course, everyone else is welcome to read and comment. You will find in this book an unconventional answer to a very large question. Perhaps it will encourage you to ask your own large questions and to seek your own unconventional answers, whatever your current phase of life.

Preface

This book redefines what it means to be human. I know it takes some audacity to make this claim, and I do not do so lightly. Many stronger intellects have offered their opinions on what humanness is. Philosophers, scientists, poets, politicians, writers, priests, ayatollahs, bartenders, cabbies, indeed in almost every occupation and persuasion, in almost every language and dialect, in almost every habitable place on the planet, virtually everyone has an opinion about who and what humans are. I would not ask you to put my views ahead of any of these others, but for one small difference in how I approached the problem. Rather than dwelling on the *Who*, the *What*, or the *How* of human nature, I focused on the *Why*. Why do human beings do what they do? Why do we build things? Why do we believe in God? Why do we paint pictures? Why do we kill each other? Why do we worry about who we are? Every time I thought I had an answer to a *Why* question, I forced myself to ask why again, over and over, until every question eventually resolved in my mind to a single root concept. This core idea I took to be the essence of humanness. It is simply this: that human nature is dominated by a propensity for *control*; that our species emerged from apedom through evolutionary enhancements in our ability to control events and their consequences; that these enhancements were sustained over time because they solved the prevailing problems of the natural world more quickly and surely than any prior or subsequent evolutionary strategy among the apes. Control is therefore the defining attribute of humanness. It is ultimately the answer to every *Why* question regarding human nature, as the balance of this book will demonstrate.

Now, you may bristle at the suggestion that human beings are controlling creatures. In common usage, the word *control* generally has a negative connotation, a tint that casts the human species in a less flattering light than we would prefer. But a moment's honest introspection should convince you that your need to control events, to manage their consequences, to shape and maintain your world in accordance with familiar and cherished personal expectations is as accurate a description of your humanness as any other label you can name. Control, as we shall define it in this book, is the root pattern that dominates human life. It is not something we simply choose to do; it is our fundamental nature, the primary operating characteristic of a unique biological machine developed over millions of years. Other species, our ape peers and contemporaries in ancient times, did not take this path, and none of them became human, at least not to the degree we did. Although they started at the same evolutionary point with the same genetic structure and potential, these other apes never adopted control as the central theme of their life histories. As a result, they have come to control very little of the world around them. We humans, in contrast, display a powerful and pervasive drive to move and shape the world to fit our internal expectations and standards. Only one evolutionary line emanating from our ancient apelike ancestors leads to a true controlling species. It is ours, and that propensity to control is what makes us human.

My goal in the following pages is to demonstrate how the specific qualities of our species, the human signature if you will, support or reflect our propensity for control. The way we think, the way we walk, our languages, our cultures, our gods, our wars, our loves, all make more sense in the context of a control strategy. In some cases, I believe, they only make sense in this context.

Understanding us is no small task. The sheer scope of this undertaking necessarily limits the analysis of some topics and themes. My apologies in advance, but keep in mind my goal here. I want to show you first that understanding our origins requires that you understand our propensity for control, and what it means in operational, biological, neurological, behavioral, and evolutionary terms. To make this point, I must focus first on the ideas and issues that best illustrate the principle. If this general thesis has merit, then over time we can start filling in the details. If it does not, that is, if you should conclude that control does not define the human species, then this work will at least have staked out a position against which productive counterarguments may be directed.

A word about the title of this book, *Homo dominus*. I chose it deliberately to nudge you out of your comfort zone about who and what we are. You may know the human species as *Homo sapiens*, after Linnaeus (1758). It translates roughly as "the Wise." Had Linnaeus focused on other aspects of our nature, besides our supposed wisdom, he might have chosen many other names. In fact, in each chapter of this book, I suggest an alternative species name (with apologies to Latin scholars). As you may have surmised, I think that the name *Homo dominus*, the Controller, fits the true human species better than *Homo sapiens*. My quest in this book is to persuade you of this view, to show you how we became *Homo dominus*, and to explore how, or whether, our path might eventually lead to *Homo sapiens*.

CHAPTER 1

Homo dominus, the Controller: The Human Strategy

Prologue

On the margins of the great African rain forest at the close of the Miocene era, there emerged a group of creatures who had a profound destiny. They were clearly apes, by any modern biological definition, but a keen observer would probably have seen some fundamental differences between them and their deep forest cousins. They lived in a different world than the rain forest apes. They saw that world differently and they behaved differently in it. In their quest for survival, these new creatures took a fundamentally different approach. In time, they acquired the power to change the world. This is their story–what created them, what made them different, why they survived.

Of course, it is also our story, yours and mine, for those ancient creatures were our ancestors. What they became is what we are, human beings, for better or worse. Paleontologists have been trying to piece together the sequence of events from them to us for many years. The story typically begins "What is humanness and where did it come from?" But the ending remains a little blurred. Clarity will probably require many more years of digging, both in the deep sediments of the earth and in the deep layers of the human psyche. Neither yields its fossils easily, and what little we find will surely spark more debate. It is quite possible that humans will be asking this same question many millennia hence, still with no resolution in sight.

We do know some things. For example, there was once a hominid species we now call *Homo erectus*. What they called themselves we shall never know. Paleontologists generally agree that *erectus* clearly marks the path toward humanness (Walker and Shipman 1996; Palmer 2006; Wade 2006). They suggest that these tall, slender creatures with the big brains and bright eyes carried the early genes of humanness expressed in ways we would probably recognize. They appear to have done many of the things we do: they organized, they made tools, they defended themselves, they took care of one another, they conversed, and they survived. Not only did they do these things in their traditional East African habitat, they also appear to be the first hominid species to venture out of Africa to claim a place in the larger world. Starting at least a million years ago, perhaps much earlier, *erectus* bands migrated to Europe, China, and Southeast Asia (Lewin 1993; Cameron and Groves 2004). Unlike their predecessors rooted firmly in the African soil, *erectus* apparently had the ability to master the outside world and the courage to take it on. In the minds of many paleontologists, such quests clearly mark the emergence of humanness (Pasternak 2004; Leakey and Lewin 1992; Tattersall 1998; Walker and Shipman 1996).

Before *erectus*, however, the human path is considerably fainter. It appears to have started modestly enough, several million years earlier, among small bands of clever bipeds scuttling between the forests and open plains (Brunet et al. 2002; Johanson, White, and Coppens 1978; Leakey et al. 1995; Senut et al. 2001; White, Suwa, and Asfaw 1994). These early hominids no longer lived in the traditional rain forest habitat of their ancestors, the Miocene apes (Begun 2003; Kelly 1992; Pilbeam 1986). That vast rich ecosystem offered food, water, and cover in abundance to any species that could seize and hold its territory. But it was no longer home to the emerging hominids. For reasons not entirely understood, our ancestors dwelled on the margins of the great rain forest in an ecosystem that presented a very different challenge.

Why the hominid ancestors left the rain forest remains a crucial question for human paleontology. One likely factor, according to paleobotanical field studies, is climate change (Potts 1996). Starting about seven to eight million years ago, global changes appear to have brought increased climatic variability to the African continent. The conditions needed to sustain the vast rain forest ecosystem became more intermittent. Moisture patterns fluctuated, the temperature range increased, and parts of the forest canopy periodically thinned and opened into patchy woodlands, and sometimes even dry savannas. These new ecosystems lasted for a time, in some cases many millennia, then the cycle reversed, and the traditional rain forest habitat returned for a time. In contrast, the ecosystem deeper in the forest interior, nearer the equator, remained more constant. Here the traditional patterns of abundance continued more or less undiminished across the millennia.

For many paleontologists, the disparities between the marginal and equatorial ecosystems represent the crucial factor in the evolutionary divergence of the apes. While the equatorial groups experienced relatively little change over the generations, ecological fluctuations were acute in the outer ranges. During times when their resources diminished, the marginal apes retreated toward the deeper forest in order to maintain their accustomed habitats.

As these immigrants crowded the indigenous populations, competition for resources increased. Eventually, the shrinking forest ecosystem could not sustain the increasing population density, and pressure developed to reduce it. If the paleontologists have it right, the apes best adapted to holding the choicest territories in the deep forest remained there, pushing the weaker groups back to the margins. The winners in this ancient competition, so the speculation goes, eventually evolved into modern chimpanzees and bonobos who remain in their equatorial territories to this day. The losers had to survive (or perish) in the sparser and more unsettled habitats on the margins.

Most paleontologists believe that human evolution began in earnest among the great apes forced to live in these marginal areas (Potts 1996; Calvin 2002; Palmer 2006; Wade 2006). Like all species, these creatures depended on their ecosystem to sustain them, but over the generations, the marginal apes experienced many different ecosystems. During drier epochs, the traditional rain forest habitat of closed foliage and filtered sunlight opened into brighter vistas over wider expanses of grass and scrub brush, leaving less shelter against heat and storms, increasing exposure to ground predators, and producing wider variations in the food and water supply over the course of the seasons. During wetter epochs, the rains returned and the woodlands and forests grew lush and thick again, restoring something closer to the ancestral habitat. This cycle repeated often, and with each turn, the marginal apes faced new adaptive challenges.

The human story might have had ended there, extinguished abruptly by a withering drought during a prolonged dry period, or a catastrophic flood during a wet phase. But something remarkable happened. In a pivotal confluence of events, evolution took a different turn, producing a new species of ape with a different strategy for dealing with its changing world. In those scattered groups of bewildered apes, our true ancestors now lost in time, natural selection began accumulating the genetic raw material that gave them the power to deal more successfully with the variability that afflicted them and their descendants. While the deep forest apes simply hunkered down and reacted to events, unaware of their causes and patterns, the marginal apes became proactive in coping with the mosaic of change presented to them. These new apes began to see the world differently, to recognize its patterns, to predict its cycles, and to act more effectively to minimize the negative consequences of change and to maximize the positive. In short, nature invented the *human strategy*, a survival repertoire that offered better solutions to the problems of life on the forest margins. This strategy embodies the concept of *control*, about which we shall have much more to say.

This book is about those ancient apes and the human genes they first carried. We carry them too, inherited from them through a long unbroken line of increasingly human creatures. They started in Africa and eventually spread to almost every habitable niche on the planet (Lewin 1993; Cameron and Groves 2004; Itzkoff 1985; Pasternak 2003). They now control a large portion of earth's resources, and may hold the key to the planet's future health. Those human genes, for better or worse, gave us greater control of our lives. The question remains whether they gave us the wisdom to control ourselves.

Framing the Question

Some critics may fault the foregoing as oversimplified. They will argue, rightly, that much of human evolution remains shrouded in mystery; that the scientific evidence, where it does exist, demands more complicated explanations. Certainly, the scientific literature on human origins occupies thousands of volumes on every conceivable element of our nature, much of it fascinating, some enlightening and compelling, some foolish and even dangerous. This proliferation continues with the almost daily appearance of new books and papers. In the end, I doubt any single account, certainly not this one, will tell the complete story of the human origin and its consequences. For the moment, that final compendium lies scattered in loose pages in offices and laboratories around the world, like the bones and broken tools of our ancestors lying in the earth just hidden from view.

Therein lies the problem. The literature on human evolution expands exponentially while our understanding follows at a more leisurely pace. The multiplicity of sources, the diversity of focus, the noise of many voices, all point to the real issue: that we really do not agree on what *humanness* means. Some see it only in bones and teeth, others in the flaking patterns of broken stones. Still others look at DNA sequences, or language syntax, or burial mounds, or cave paintings to find the first real humans. The list goes on. The proliferation of possible solutions reflects only the vagaries of the problem. No single answer suffices because we have not asked the proper question. If we cannot state simply and clearly what makes us human, in the biological sense, then our effort to trace the path of human evolution, which operates *only* in the biological sense, inevitably goes adrift in a sea of confusion and contradiction.

So my prologue did not intend to provide a definitive account of human origins, but rather to set up a better question. Suppose we take an imaginary walk along earth's biological timeline and stop at the period of climate change mentioned above. We look back and see a long trail labeled "No Humans here." We look ahead and see a much shorter trail labeled "Humans here." Something important happened at that point to create this new thing called a human being. It seems to have happened relatively suddenly, as if some nonlinearity warped the course of biological history. That moment defines the start of the human strategy. From that point on, the world had to deal with a species that did not simply hunker down and endure nature's indifferent assaults, but instead took control of events: building, regulating, and transforming the world to suit its needs. That path led ultimately to the species I call *Homo dominus*.

Homo dominus Overview

Yes, I said *Homo dominus*, not *Homo sapiens*. Our traditional species name, *sapiens*, comes from the great taxonomic system of Linnaeus (1758). Latin dictionaries variously translate it as "wise, sensible, judicious." Whether Linnaeus truly believed our species possessed those traits, or whether he simply gave us the name as a hope (or a prayer), we do not know. The relevant question is whether the name *sapiens* fits the human species. I remain skeptical. It might one day, if we behave wisely, sensibly, and judiciously. But even in Linnaeus's time, the *sapient* species engaged in merciless warfare, slavery, genocide, torture, exploitation, pollution, pornography, and idolatry, all on a scale that dwarfed anything displayed by other species. Such behavior probably dates well back into our past, and a quick scan of the morning news suggests it continues today. On what basis, then, do we call ourselves wise, sensible, and judicious?

In my opinion, the name *Homo dominus*, the Controller, fits us much better. Far from taking the enlightened path we would like to imagine, the human species survives on a diet of control. It is a deep-seated drive that often ignores its negative impacts on other species, ecosystems, and even other human beings. Starting very early in our evolutionary history, we perceived the world differently and took action to bring those perceptions into line with our growing expectations and needs. Control therefore distinguishes our species from others, not wisdom, sensibility, or judiciousness. In that sense, control defines humanness.

This human strategy contrasts with that of our close genetic cousins, the chimpanzees and bonobos (Goodall 1996). They exhibit a survival strategy we might characterize as *reactive opportunism*. They move patiently through their range looking for fruit or attacking hapless monkeys. They form complex societies, competing with other groups for parcels of territory. Some use tools and occasionally walk upright. Mostly, however, they simply wait patiently for opportunities to exploit. They seem unable or unwilling to control their world, to comprehend its causalities, to project its consequences, or to deal proactively with the changes that inevitably come. Chimpanzees and bonobos have large brains, as primates go, and laboratory studies have shown that they clearly have the capability to do much more than they actually do in the wild. There, they seem to underachieve. Their survival strategy appears to focus simply on holding the bounty of the rain forest against other apes, pushing weaker groups out of the choicest territories. They have pursued this course for millions of years. Success has given them possession of an incredibly rich habitat that meets all their needs, but whose limits and future they do not comprehend.

Of course, now the tables have turned. If the ancient struggle had the deep forest apes driving out the marginal apes, the modern version has the marginal apes returning with a vengeance to conquer the forest apes. Increasingly, chimpanzees and bonobos face the loss of their habitat to human encroachment. Most populations now rely on rare human goodwill to protect them against other humans. In time, as their range disappears, they will either go extinct, or live a sheltered existence in zoos and preserves as a curiosity for human children and research material for comparative biologists. It appears their survival strategy of holding the rain forest has finally failed. In contrast, the human species has expanded to nearly every habitable niche on the planet, plus occasional forays off-planet.

Paleontologists believe that humans and chimpanzees started at the same evolutionary point about six to seven million years ago, in a time of increasing climatic and ecological fluctuation on the African continent. In this span, a mere blink in geologic time, natural selection has produced two genetically related but hugely different species. One endured in a closed forest world, the other took control of an open hostile world. Understanding this remarkable divergence remains one of the central problems in human studies.

The Nature of Control

To understand *Homo dominus*, you must understand control as a biological process. This may be new to some readers, so I will take some time to define it in detail. Although control theory has been around for half a century, its importance in human evolution remains unclear. Common dictionary definitions of the term *control* include phrases such as "to exercise restraint or direction over; dominate; command." While this certainly describes general tendencies of human nature, the application of control theory to human evolution requires greater biological specificity. Fortunately, modern theorists have provided us some better ways of looking at the problem (Cziko 2000; Marken 1992; Powers 1973; Rosenbrock 1990).

In the present context, the term control refers to *action taken to bring perception into line with expectation*. We shall refer to this definition repeatedly in the following chapters, so it is important that you understand it. It connects three fundamental elements:

- Action
- Perception
- Expectation

Moreover, it connects them in a very specific way, as shown in Figure 1.1. This arrangement provides a major key to the human puzzle, so let's look at it in more detail.

First, the concept of *perception* should be familiar to all of us. It refers to patterns of information flowing into an organism from its sensory systems. Control theorists generally refer to this as *input*. All advanced biological organisms have arrays of sensory elements, called transducers, which pick up bits of information from their local environments and transmit them into the nervous system. Here they combine to produce the neural correlates of the external world. Although perception usually refers to information coming from the outside world, it can also apply to internally generated input patterns. Such internal perceptions lie within the realm of the imagination. They are often as important as external perceptions for understanding human behavior.

Action is also a familiar concept. It refers to the things an individual does: making movements, exerting forces, secreting chemicals, vocalizing, and so on. We usually think of action as overt behavior that an outsider might observe and measure. However, it can also include internal activity in the body's organs and tissues, like the brain or the adrenal cortex, which an outsider cannot observe directly. Control theorists use the term *output* to refer generally to the action part of a control system.

Expectation is a trickier concept that requires more explanation. Control theorists use terms such as *reference, specification*, or *standard* for this element. By expectation, I mean an internal pattern of neural activity against which the system compares perceptual activity. In other words, an expectation is a kind of neural template that screens perceptual information. When a particular expectation is active in a given nervous system at a given moment, it defines the perceptions that the individual, as a control system, requires or seeks at that moment.

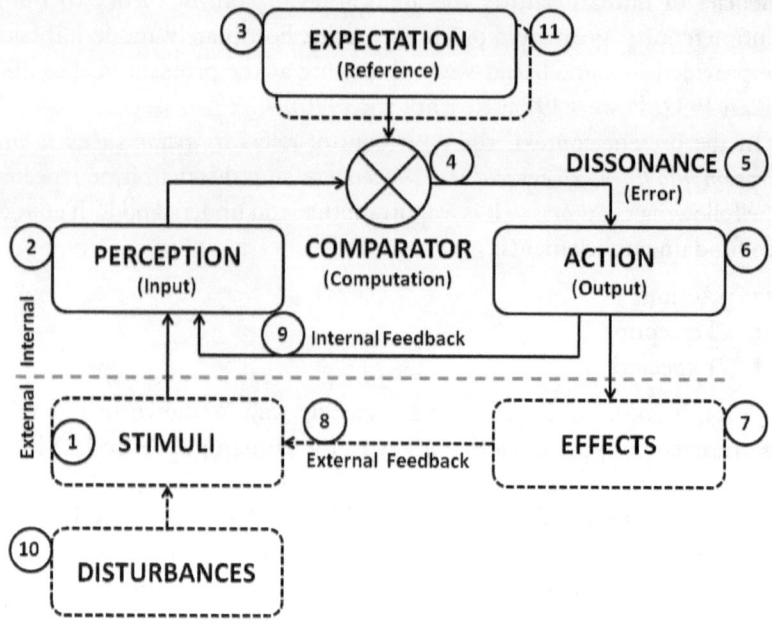

Figure 1.1. The biological definition of control relates three fundamental elements—perception, action, and expectation—in a precise relationship. Control theory uses the terms input, output, and reference, respectively. The control architecture depicted here defines a recurrent signaling loop between perception and action, modulated by internal expectations and external disturbances. See text for details on numbered items.

The term expectation, in the present context, should not be confused with the term *expectancy* used by early behavioral theorists (Tolman 1932). Although there are some conceptual similarities, and those researchers were trying to solve some of the same problems, the two terms are not strictly interchangeable.

Physically, expectations comprise neural activity patterns built up from many sources. Fundamental biological drives, like hunger, thirst, or sexual arousal, represent one major source. For example, when I am hungry, my nervous system instinctively develops activity patterns that represent basic expectations about food. Once they become active, my nervous system, acting as a control system, compares all incoming perceptions against them. In this particular screening process, only food-related perceptions generate a match. Thus, steak, salad, and cookies fit my food expectations, while rocks, tables, and lamps do not. As long as I am hungry, and as long as food-related perceptions do not present themselves, I will keep searching, and my control system will keep screening. Similar rules apply to all other basic biological drives.

The other major source of neural expectations is the set of learned events. These are neural patterns, encoded in memory, that reflect past events in an individual's life. Examples include learning how to find and prepare food, navigating a safe path through the forest, or, in modern life, driving a car. Learned expectations work similarly to the instinctive ones. For example, when I drive a car on an American two-lane road, I carry an expectation, learned from previous experience, that my vehicle should remain in the right-hand lane. As my senses transmit perceptual information about the position of the car, my internal control system constantly compares my perceptions against that learned expectation. British drivers, of course, learn to expect their vehicles to remain in the left-hand lane. None of us is born knowing which lane to drive in. It is something a modern human learns, along with a vast collection of other expectations.

Control theory defines a precise interactive flow among expectation, perception and action, as shown in Figure 1.1. Note the critical feature of *circularity*. Information in control systems flows in loops from input to output and back. Various factors may modulate this flow, but the essential nature of control is cyclic activity in a looped, or *recurrent*, structure. It is important that you understand the circularity of control, as we will use it later to explain some important human characteristics.

Let's take a turn around the basic control loop in Figure 1.1. First, note the line separating *external* from *internal*. Some events in the control loop may take place outside the individual's body in the external world, while others take place inside. Internal events refer to the collective interactions among neurons, muscles, organs, and tissues that comprise the biological organism. External events refer to the collective interactions in the physical world of matter and energy outside the confines of the organism. Of course, an event that appears external to one organism might be internal to another. The internal-external line is simply an arbitrary divider drawn from the frame of reference of a particular control system.

Starting on the left side of Figure 1.1, external events, or stimuli (*1*), trigger internal neural activity that forms the elemental basis of perception (*2*). As this activity flows through the nervous system, it interacts with other neural systems that encode sets of expectations (*3*), that is, neural patterns already active in the system. Control theory says that this step is essentially a mathematical computation that has the logical attributes of a *comparison* operation (*4*). In other words, the system compares external representations of the world to internal ones. The result of this neural computation is a new pattern of neural activity we call *dissonance*, or *error* in control theory

9

(5). Dissonance in this context is simply another set of neural activities whose pattern and intensity vary according to the outcome of the comparison between perception and expectation. For simplicity, control theorists say that dissonance (error) is low whenever perception (input) and expectation (reference) match and high whenever they do not. Thus, the term dissonance is simply a measure of the degree to which incoming perceptual information matches what the system expects.

In the next stage, the dissonance activity flows into the action system (6), where it can trigger various behaviors: locomotion, vocalization, orienting, and so on. Precisely which actions occur in response to particular dissonance patterns depends on many factors, including the individual's instinctive repertoire, prior experiences, and the existing context. In some cases, certain dissonance patterns will produce well-defined action patterns, and the individual will act in the smooth, coordinated fashion, characteristic of a well-established control situation. In other situations, the control system may have no preset responses to particular dissonance patterns, and the resulting actions may appear random and uncoordinated.

Action, whether coordinated or random, produces *effects*, as shown in the next stage (7). Effects are the external consequences of action. They may be quite varied. For example, the individual may physically relocate to a new point in space, reorient to a new compass heading, move eyes or limbs, shift the position of objects, emit sounds, secrete chemicals, and so on. As effects propagate through the external milieu, they alter the environment, which in turn alters the original input pattern. This is the crucial point: *Effects change perceptions.* Thus, external events, whether large or small, represent potential points of closure in the control loop. As new and altered perceptions sweep back into the system (1), the control loop immediately restarts and the entire process repeats. For example, an action as simple as the movement of the eyes changes the visual focus, bringing a new pattern of visual inputs to the retina. This new perceptual mix flows into the system, combines with existing expectations, which produces new dissonance patterns, which trigger new actions, which produce new effects, which again alter perceptions, and so on. The process repeats over and over.

In addition to external feedback through the environment, actions may also produce internal effects in the form of direct neural activity from the action centers back into the perceptual areas of the brain (9). This internal feedback can also alter perception, which immediately restarts the control loop. Internal feedback represents an important element of human control processes. We shall have much more to say about it later.

This seamless circular interplay of perception, action, and expectation continues indefinitely until a critical event occurs: specifically, whenever perception *matches* expectation. This happens whenever the incoming perceptual patterns precisely null the ongoing expectation patterns. When this happens, the comparison operation (4) produces significantly less dissonance, which in turn causes action generated in that particular module to abate. In effect, this control system has found its desired perceptual endpoint, as defined by its current expectation, so it requires no further action. The system relaxes into a kind of null state, where it remains, unless and until subsequent events restart it.

To illustrate this process, let's consider a simple example from modern human life: a person driving a car. Successful driving requires the driver to maintain the position of the vehicle between designated lane markers (expectation). To accomplish this, the driver must perceive the actual position of the vehicle relative to the markers (perception). If the perceived position does not match the expected position (dissonance), the driver must move the steering wheel in the appropriate direction (action). This movement changes the angle of the tires which forces the car back into the lane (effect), creating a new perception of the car's position relative to the lane markers. If the driver's new perception matches expectation (null state), the wheel-turning behavior ceases. If not, the driver must continue to move the steering wheel. With an experienced driver, this control loop functions smoothly, continuously, and seemingly effortlessly. As long as the vehicle stays in its lane, the driver need take no further action.

Modulation of this control loop may occur in two ways, as shown in Figure 1.1. First, random external events, called *disturbances* (*10*), may disrupt the flow. Suppose a gust of wind hits the car and moves it unexpectedly from its lane. The driver, detecting the sudden mismatch between perception and expectation, turns the wheel appropriately to restore the desired perception. The disturbance momentarily elevates dissonance; the steering reaction quickly nulls it and restores equilibrium. Such disturbances are inherent in a world of fluctuations, both external and internal to the organism. Energy and matter are in constant flux, due partly to physical processes and partly to the actions of other creatures, both human and nonhuman, with whom we share the environment. Such events constantly disrupt established control balances, producing constant flickering dissonance that drives actions to restore and maintain the desired equilibriums.

Control can also be disrupted by internal changes in expectation (*11*). This is a more complicated process. In our driving example, suppose the driver needs to turn onto another road at some point in order to reach a desired destination. If the driver's control system holds rigidly to its initial expectation, it will perpetuate the initial course and the driver will fail to make the turn. So, as the vehicle nears the critical point, the driver's internal control systems must somehow be reprogrammed with a new expectation pattern reflecting the new direction. As the new expectation takes hold, it creates an immediate mismatch with existing perceptions. In other words, where the driver's current perception was consonant with the prior expectation, it is dissonant with the new one. This new dissonance pattern motivates new action. The driver again moves the steering wheel in the appropriate direction. As the vehicle comes around to the new heading, the driver's perception changes again, this time to match the new expectation. At this point, dissonance again declines; the turning behavior ceases; and the system reverts to its previous straight-line control pattern, but now traveling on a new course heading. Note that the world did not fluctuate to produce this new behavior. It came from an internal change in the neural activity encoding expectations. By executing programmed sequences of such expectations, control systems can execute seemingly purposeful behaviors of great complexity, intelligence, and insight. Mechanistically, however, they are simply repetitions of the same basic control process. From that perspective, such programmed changes in internal expectations, whatever their origin, are functionally equivalent to the effects of external disturbances. Both produce an immediate uptick in dissonance, which in turn motivates changes in action.

Control thus reflects the reciprocal interplay between perception and action, modulated by variable expectations and disturbances. Behavior, therefore, both depends on perceived stimuli and determines what stimuli will be perceived. The environment alone does not dictate an individual's behavior, but neither does the individual behave independently of the environment. In the control model, behavior, in all its forms and complexities, simply reflects the ongoing sets of matches and mismatches between perception and expectation resident in the behaving system.

Now, some theorists will argue, rightly, that the above analysis virtually guarantees that every biological system displays control behavior to some degree (Odling-Smee, Laland, and Feldman 2003). And if we are defining humanness in terms of control, we must necessarily attribute some degree of humanness to other species whenever they exhibit comparable control behavior. But, of course, we do. Whenever we witness control behavior in other species, we tend to attribute the same characteristics of intelligence, purpose, and self-awareness to them, as we would otherwise reserve for the actions of our own species. Consider a pride of lions cooperating to maneuver a hapless gazelle into an ambush, or a beaver constructing a dam, or a chimpanzee using sign language to secure a treat. Even robotic machines take on anthropomorphic characteristics when they purposely, and seemingly intelligently, adjust their outputs and inputs to achieve a clearly defined outcome. Adding a smiling face and a voice synthesizer to such machines enhances the illusion, if indeed we should even call it an illusion. The patterns of control, even when exhibited by nonhuman entities, invariably elicit some degree of anthropomorphization by human observers.

The real difference between *Homo dominus* and the rest of the animal kingdom lies not so much in the quality of its control behavior, but in its scale and scope. While other species control only limited aspects of their worlds, the human species controls with much greater reach and impact. A moment's reflection should convince you that this is so. Do we not carry a rich and detailed set of expectations about how things are supposed to be? Do we not spend most of our waking day trying to bring real events into line with these expectations? It all happens so naturally and instinctively that we rarely recognize it for what it is: a deep-rooted biological predisposition that we find nearly impossible to moderate. Where other species exhibit modest and selective control behavior, the human species displays it boldly and pervasively. This fundamental shift propelled us out of apedom, as we shall now explore.

Control and the Quest for Stability

If the ancestral apes on the margins of the great African rain forest were trapped in an ecological oscillation driven by a fluctuating climate, then the emergence of humanness could perhaps be characterized as a quest for stability. And if the shift toward humanness was a quest for stability, then the notion of *control* as the defining characteristic of humanness begins to make sense. Control is, by definition, a process of stabilization. Control arrangements are always potential solutions when the problem is instability. Thus, the selective augmentation of control behavior among the marginal apes of the late Miocene would seem a logical response to their evolutionary predicament. Faced with an ever-changing habitat, alternating between rain forest and open woodlands, wet and dry, equatorial and temperate, with the necessities of life, such as food and water, and the threats to life, such as predation and disease, all in constant flux, the marginal apes would have been subjected to powerful selective pressures that favored the emergence of improved stabilization mechanisms to cope with their changing world. In contrast, the apes of the deep forest, closer to the equator and inland from the ancient oceans, suffered less dramatic fluctuations and, as a result, may have experienced less selective pressure toward radical control enhancements. Thus, the key question for understanding human evolution becomes how the fluctuating habitat on the forest margins triggered an evolutionary shift toward control behavior (i.e., humanness), while the more constant environment in the forest interior preserved the more apelike forms. Some ideas on this critical differentiation follow.

Classical Darwinian Theory (Darwin 1859; Darwin and Wallace 1858; Wallace 1869) holds that in any species population, variations constantly accumulate in the genome, each producing slightly different characteristics at the molecular, physiological, behavioral, and social levels. Each genetic variant in a population represents a biological test of reproductive fitness under the prevailing environmental conditions. The reproductive success of any particular variant indicates a relatively favorable adaptation, while reproductive failure indicates the opposite. So it was with the late Miocene apes. Whenever the prevailing environment featured a slow warming trend, ape variants better acclimated to warmer, wetter conditions tended to survive and reproduce more successfully than their less warm-adapted peers. Such genetic variants might have included biochemical variations that improved thermoregulation, metabolic mutations that that conveyed better utilization of food and water, or behavior modifications that proved more conducive to survival and reproduction under hotter, wetter conditions. Conversely, during a slow cooling trend, ape variants better adapted to cooler, drier ecological niches had a reproductive advantage. As their reproductive success increased their numbers, these variants gradually transformed the indigenous population into an evolved form better suited to a cooler, drier ecology.

According to the classical Darwinian view, as long as the rate of genetic diversification in a breeding population keeps pace with the rate of environmental change, an evolving line can persevere indefinitely. However, if environmental change happens too quickly, or genetic diversification too slowly, then the species runs a greater risk of extinction. Darwin himself argued for slow incremental change as the primary characteristic of evolution. Modern theory, however, suggests that the pace of evolution may change dramatically, depending on the pace of global events (Eldredge and Gould 1972; Berggren and Van Couvering 1984). A sudden cataclysm, such as the now famous Chicxulub asteroid strike some sixty-five million years ago, can rapidly delete large numbers of species, such as the dinosaurs of the late Cretaceous period. Had the pace of change been slower during that period, dinosaurs might still be roaming the earth in recognizable form.

Although the scale and pace of climatic change in the late Miocene were undoubtedly less dramatic than those of the late Cretaceous, the climatic instability facing the ancestral apes was certainly sufficient to threaten them with extinction. Moreover, two additional factors complicated the apes' situation.

The first was the great apes' relatively low reproductive rates. Modern species average about thirteen to eighteen years per generation. We assume that their ancestral forms had roughly comparable reproductive cycles, perhaps varying by a few years more or less. The problem is that longer generation times result in a slower spread of genetic diversity into a population, leaving a species more vulnerable to rapid ecological change. In a slowly reproducing species, a potentially beneficial mutation takes longer to replicate itself than in a faster reproducing species. Even though both species require the same number of reproductive pairings across the same number of generations to achieve the same relative gene frequency, the slower reproducing species takes more total time to reach this level. If the pace of environmental change quickens, fewer members of the slower reproducing species will have acquired the desirable mutation, and a greater proportion of the population will therefore remain vulnerable.

The second problem for the marginal apes was the apparent cyclic pattern of ecological change (Potts 1996). The marginal habitats did not go through long-term, monotonic change, but rather through a series of alternating climatic reversals of relatively shorter duration. Arguably, oscillation represents a harsher evolutionary challenge than monotonic change. During any particular phase of the cycle, selection tends to favor genetic variants optimized to the immediate trend. Thus, during a warming phase, the population accumulates warm-adapted variants and sheds cold-adapted ones. As the trend reverses, however, the selection pressures also reverse. If the trend reverses yet again, the evolutionary forces shift yet again, and so on. In oscillating environments, genetic changes accumulated during the preceding phase of the cycle must be reversed or masked during the subsequent phase. Each phase must *undo* the work of the previous one. Over a period of, say, one million years, a climate oscillation with a period of fifty thousand years would complete about twenty cycles, each covering about three thousand typical ape generations, assuming about seventeen years per generation. Depending on the mutation rates of the relevant genes, an ape species might keep genetic pace with this rate of oscillation according to classic mutation/reversion dynamics. However, if the rate of environmental oscillation ever exceeded the rate of genetic variation, then the slowly reproducing apes would necessarily lag in their race to adapt, leaving them at greater risk of extinction. In short, the climatic oscillations of the late Miocene presented an especially difficult evolutionary challenge to the ancestral ape populations. The unique selective pressures produced under these conditions led to the fundamental divergence of the hominids from the ancestral apes, as we shall now see.

The Fundamental Divergence

Logically, if the problem is chronic environmental instability acting on a slowly reproducing population, then natural selection should tend to favor the emergence of stabilizing mechanisms with shorter time constants than the classic mutation/reversion approach. Rather than running a constant, and probably futile, genetic race against reversing climatic trends, a marginal species might find a more enduring solution in a general set of coping mechanisms that moderate the consequences of change itself, regardless of its direction. For example, in an oscillating climate, a species with the ability to cope with *both* warming and cooling trends should have a reproductive advantage over a species optimized to one or the other. While the latter might gain during a particular phase of the climatic cycle, it would tend to fall back sharply during subsequent reversals. The flexible species, in contrast, might maintain population growth during both phases of the cycle, albeit perhaps at a suboptimal rate for any particular phase. Over a prolonged period of climatic oscillation, the species with the more flexible coping strategy might eventually win out over its boom-or-bust peers.

The key question, of course, is what mechanisms could convey such flexibility? More specifically, what inheritable genetic changes could yield a phenotype of relative immunity to climatic oscillation? One possible answer is greater *behavioral* flexibility. Rather than accumulating physiological or anatomical changes optimized to particular trends, a species might accumulate gene variations that support behavioral coping strategies with much shorter time constants and much wider adaptability to various environments. Compared to their peers, such behavioral mutants might not have optimal adaptation to any particular climatic condition. However, they might have a significant advantage in their quicker acquisition and expression of short-term coping behavior in any given climatic condition. Such behavioral flexibility would tend to moderate the consequences of change, regardless of direction. Thus, in a period of warm, wet conditions, these behavioral mutants could adopt a repertoire of short-term actions that would specifically counter the negative effects of warm, wet conditions. During a subsequent cooling, drying trend, these same genetic lines could elaborate new behavioral patterns that reflected the new realities of cooler, drier environments. Physiologically, they were not optimized to either warm wet climates or cool dry ones, but behaviorally, they could cope with both, at least at a level sufficient to preserve a viable reproductive base. Perpetuated over many climatic cycles, this rapid response strategy could have preserved an underlying genetic legacy that gradually differentiated the marginal apes from their deep forest cousins, eventually producing recognizable hominids.

Now, if the principle of internal stabilization against external change represents the defining marker on the path toward humanness, then the crucial question becomes what kind of behavioral adaptation best delivers on the promise of rapid and flexible stabilization. The answer, of course, is control behavior, as we have defined it above. The following section outlines the fundamental tenets of this hypothesis.

Control, Stability, and Human Evolution

In two of the three basic elements of control, perception and action, humans do not appear to have disproportionate advantages over other species. Many species have better perceptual tools, such as keener hearing, vision, and olfaction. Others have much more impressive motor skills, such as speed, strength, and agility. Indeed, our species is actually relatively deficient in some of these areas, compared to other species. Thus, the fundamental source of human divergence, from a control perspective, must lie in the nature and evolution of the *expectation* component. It is here that we must focus our attention.

Let's start with some simple operational examples of what we mean by expectation. Consider the electronic feedback circuit shown in Figure 1.2A. Such circuits are familiar to every first year electrical engineering student. The control principles embedded in this simple design are identical to those in Figure 1.1. This circuit creates a stable interaction between input and output in order to achieve a desired performance characteristic (i.e., an expectation). In this case, the source of the expectation is the engineer who designs and builds the circuit. If the engineer wants a particular voltage at any particular point of the circuit, she can select and arrange the interacting to create just such a result. By using a feedback control circuit, she guarantees that the system will resist random fluctuations and disturbances, from both outside and inside the components, and will remain stable. A successful design means the circuit performs in accordance with the engineer's imposed expectation.

Now let's extend this basic idea to biological control systems, as suggested in the hypothetical neural circuit shown in Figure 1.2B. The principles here are the same, even though the nature of the components is very different: protein, carbohydrate, and nucleic acids versus copper and silicon. Regardless of their material structure, however, the same control principles apply, provided both arrangements exhibit the same interactions. Just as the electronic circuit achieves a stable null point, so too does this neural circuit. The essential property of stability is also present: any shift in the input neural activity will cause the neural feedback circuit to adjust its activity to stabilize the target activity.

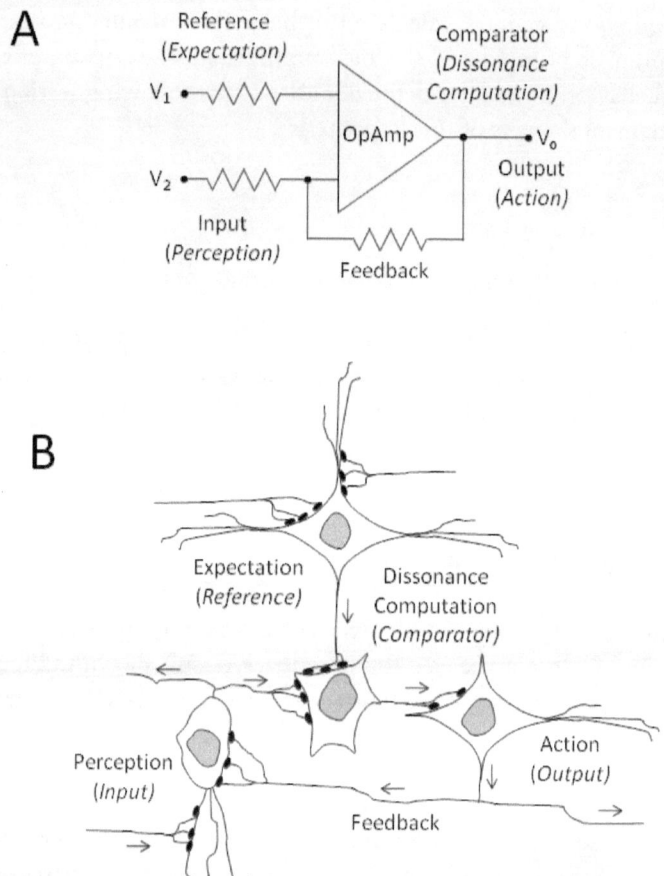

Figure 1.2. Two examples of control architectures using very different substrates. (A) A well-known operational amplifier circuit that compares the input signal with a reference signal, then feeds the output back to modulate the input. Some alternative *cognitive* labels are shown. (B) A hypothetical neural circuit that incorporates the same signaling relationships as the electronic circuit. Even though these components are biologic, the same control principles apply. The cognitive labels look more at home in this version.

What differs in the biological context is that the term *expectation* is a little harder to explain. In electronics, the notion of a design engineer imposing his expectations seems completely natural. In biology, the notion of a Design

Engineer imposing His expectations on living systems has obvious theological implications, which we will not pursue here. Absent such supernatural forces, the only abiding principle for defining biological expectations is Darwinian selection. Thus, the concept of expectation in Figure 1.2B lies in a particular set of neural activities that convey some reproductive advantage to the individual. With successful reproduction, those same brain structures may perpetuate through subsequent generations, and their characteristic patterns of neural activity may represent the same basic expectations in subsequent generations. Biological expectations, therefore, are simply patterns of neural activity whose adaptive values follow basic Darwinian logic. Like all else biological, they originate as the products of genetic diversity spun out by mutation and recombination, tested by reproductive success or failure among the resulting phenotypes, and therefore validated or failed according to the evolutionary challenges faced by the species.

For the marginal apes of the late Miocene, chronic ecological instability provided a powerful selective force that favored, I believe, the natural selection of genetic variants with an augmented neural capacity for representing expectations. Initially, this augmentation probably appeared as an overlay on an otherwise apelike control system. In time, however, it led to a fundamental shift in how the emerging species dealt with the world. Put simply, hominid evolution began with a richer internal representation not of how the world *was*, but of how it *should be*.

Figure 1.3 depicts this critical event in simplified form. It shows first a basic control system in the prototypical species. This species has a normal expectation represented by the box labeled *holding a banana*, a food favored by the ancestral apes, we assume. Let's further assume, primarily for the sake of narrative focus rather than biological reality, that this is the only food expectation the normal species has. As we outlined above, whenever this expectation is active, the animal will attempt to bring its perceptions into line with it. As a control system, *it cannot do otherwise.* Thus, the animal necessarily engages in behaviors such as searching, orienting, locating, approaching, and grasping bananas. Once the animal is in fact *holding a banana*, the match is perfect, perception has converged on expectation, and no further action is necessary. Conversely, as long as the animal is *not* holding a banana, perception does not match expectation, dissonance remains high, and action *must* therefore follow until the animal perceives a banana in its grasp. If bananas are plentiful in the local area, and reachable with reasonable effort, the normal control system depicted in Figure 1.3 represents a simple adaptive mechanism for acquiring them.

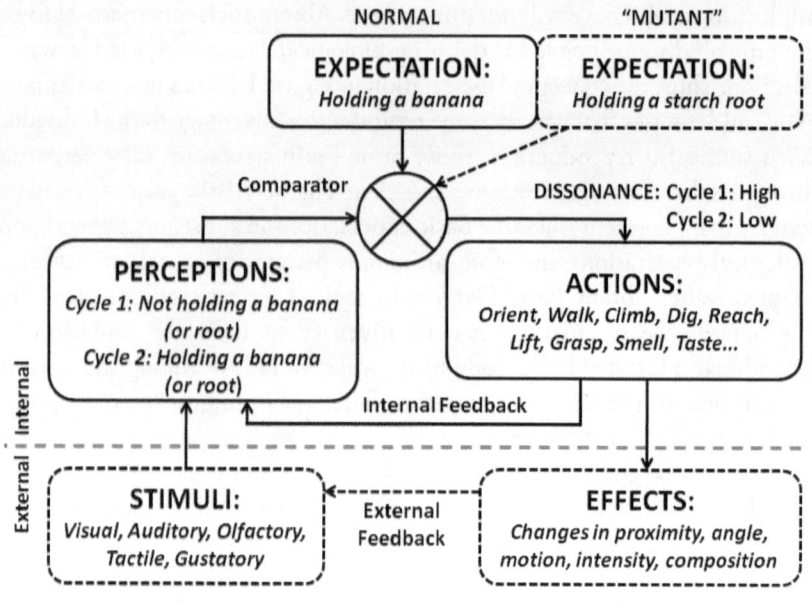

Figure 1.3. The hypothetical critical event in the hominid divergence was the expansion of neural systems in the brain that mediate the Expectation (Reference) function in control architectures. Here we see the addition of a second Expectation to the original, or *normal*, configuration. The *mutant* species therefore has two food-related expectations: *holding a banana* and *holding a starchy root*. This functional expansion requires an incremental increase in brain tissue. As a result, the mutated variant gains behavioral flexibility to deal more effectively with environmental variations in the food supply. This may prove more adaptive under conditions of ecological instability.

What happens, however, if bananas are not plentiful or reachable? Suppose, for example, a local drought or blight causes the indigenous banana trees to die off or yield less fruit. In this new environment, maintaining an exclusive expectation pattern for bananas, something that is no longer easy to find or reach, represents a behavioral dead end. The consequence may be the starvation of the local population and the potential extinction of the species. Thus, the limited control repertoire depicted in the *normal* case in Figure 1.3 is adaptive only in banana-rich habitats, where the individual's single-minded expectation always yields a reasonable probability of success. If environmental change renders that expectation obsolete, the animal's inflexibility becomes a liability.

Now let's augment the normal system with a *mutation*, as shown in Figure 1.3. This simple change represents the pivotal shift in neural control structures that nudged the ancestral apes onto the hominid path. The essential difference between the normal and mutant types is that the mutant variety has a slightly *expanded* capacity for forming expectations. In this example, I have added the expectation of *holding a starchy root*, again assuming that starchy roots were a potential food source for the ancestral apes. For the sake of argument, let's further assume that in this local habitat, bananas grow in abundance during warm, wet climatic conditions while starchy roots are more prevalent during cooler, drier periods.

With this added expectation, the control system in Figure 1.3 displays greater flexibility. Either expectation card may be active at any given time. This gives the hungry ape two potential behavioral repertoires for securing food. When the banana pattern is active, the animal behaves in ways that bring bananas into its grasp. When the starchy root pattern is active, the animal behaves in ways appropriate to perceiving and acquiring starchy roots. The control process is the same in both cases. The animal engages in the relevant actions, based on the specific mismatch between perception and expectation. The action succeeds only when the desired food is in the animal's grasp. At that point, when perception matches expectation, the control system settles into a stable null state, and the animal's food-seeking behavior ceases.

Note that we need assume no other differences between the two species, other than their differing capacities for expectation. Both species may retain the same basic perceptual capabilities (vision, hearing, touch, taste, smell), both may have the same basic motor capabilities (muscles, tendons, bones), and both may have the same internal organs with the same metabolic and physiological characteristics. Both could subsist on either bananas or starchy roots, but only the mutant form has the built-in behavioral propensity to search for and acquire *both* kinds of food. Physically, this means simply an additional increment of neural tissue, in the mutant brain, that generates an additional food expectation pattern.

Now, if both species lived in an environment where bananas were abundant and easy to reach, the adaptive advantage of the second expectation pattern might never become an evolutionary factor. Even though the variant species has two potential food-seeking repertoires, its actual behavior is just as rigidly determined as in the original species, once a particular expectation becomes active. Thus, in an environment rich in bananas, the behavior of the two species is essentially indistinguishable.

Suppose, however, that the supply of bananas declines, as suggested above. Now the behavioral inflexibility of the original species is a problem. The variant form, however, has a second expectation pattern, the one for starchy roots, which supports an entirely different food-seeking repertoire. As the mutant brain switches to this expectation, the variant species immediately switches to the root-seeking repertoire. In contrast, the normal species continues its traditional, but now increasingly futile, banana-seeking pattern. If the environment continues to change, and the supply of starchy roots eventually exceeds the supply of bananas, the variant ape may thrive, while the original declines. Depending on how fast the respective food supplies change, the replacement one by the other may take a few thousand generations, or just a few.

What is important here is not that one species replaces another, or that one food preference replaces another. We could theoretically achieve the same result simply by mutating the single expectation card in the original species from *holding a banana* to *holding a starchy root*. This solves the problem equally well, as long as the shift in food supply occurs sufficiently slowly to allow the starchy root mutation to diffuse widely enough into the population. What really matters is how these two hypothetical species would fare in a rapidly oscillating environment, where conditions alternate repeatedly between those favoring bananas and those favoring starchy roots. For the original species, the entire process of mutating and replacing the single expectation card must be repeated over and over, each cycle reversing the previous one. For the new variant, however, *both* expectation cards remain present and capable of expression in any given environment. By carrying the neural capacity for *both*, even though one may not be needed or relevant during some periods, the mutant species gains a hedge against future oscillations in its food supply.

Imagine now the proliferation of such additional expectations accumulating steadily over the generations as altered genes continue to augment neural control structures (see Figure 1.4). With each incremental addition of neural tissue, the mutant form acquires an incrementally greater range of expectations about the surrounding world. While any one expectation might be relevant at any one point in time, most remain irrelevant at any actual point in time. But by carrying these extra *dormant* expectations, albeit perhaps at some significant biologic cost (more on this later), the flexible species improves its survival potential across a wider variety of possible environments. When change comes, particularly rapid change, the flexible species adapts more rapidly, and therefore has a greater potential to endure and reproduce. The behaviorally rigid species, although physically and metabolically capable of doing all the same things, simply does not think to do them. Instead, they cling rigidly to a smaller set of expectations, which produce a more limited behavioral repertoire. This failure puts them at risk.

Although this analysis is undoubtedly oversimplified, it illustrates how a subtle change in brain function, as it relates to control systems, could have produced a significant evolutionary shift that appears to fit the hypothesized conditions of the late Miocene. For a given set of perceptual and motor capabilities, the hypothetical variant carrying a slightly larger deck of potentially relevant expectations instantly acquired a more flexible repertoire of potentially adaptive behaviors. These variants were more capable of coping with rapid environmental change, compared with their more limited peers. As the latter edged steadily toward extinction, the control-enhanced apes gradually became the predominant species on the forest margins. With each successful generation, the genetic changes that underlay their neural augmentations increased in the population. Although the variant apes probably also suffered during extreme climatic swings, they generally fared better across a wider range of environmental fluctuations than their peers. This was the first crucial step on the path toward humanness.

The Human Signature

If the theory of *Homo dominus* is correct, then the fundamental divergence of the hominid line began with incremental growth in the brain structures that mediate the expectation component of neural control systems. Logically, then, it follows that all the unique features and traits of the human species, as we know it now, must relate to this fundamental shift, either directly or indirectly. In other words, as the propensity for flexible control gained prevalence in the earliest hominids, it precipitated the distinctive cascade of physical and behavioral changes that accelerated our biological divergence from our ape cousins. This entire constellation of human traits, which I call the human signature, traces back to the initial shift hypothesized above. The balance of this book systematically explores these links.

As we shall see in subsequent chapters, some elements of the human signature are clear and direct reflections of control as a basic survival strategy. Others, however, are more subtle and indirect. Control, like any other biological process, has the potential to benefit a species under certain conditions, but also to destroy it under other conditions. In the human case, control vaulted our species into a position of dominance among the apes. However, left unchecked, control also had the potential to put our species on a new path to extinction. The human signature is therefore a complex mix both of direct manifestations of control and of indirect restraints. Without the former, our species would be little different from modern chimpanzees. However, without the latter, the human experiment might have slipped quickly and brutally into extinction.

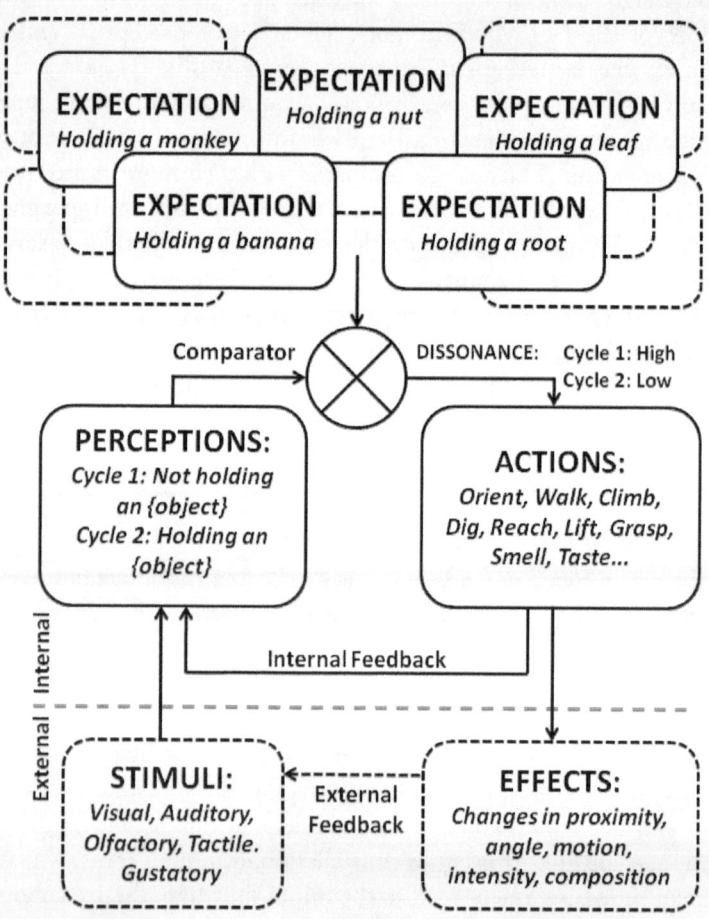

Figure 1.4. The further expansion of the hominid brain enabled a significant enrichment in the range of expectations available to a given control architecture. The activation of any of these Expectation cards, either alone, in sequence, or in combination, supports diversified and flexible behavioral repertoires that gave the hominids a better chance to cope with their challenging environment. Compared to their ape cousins (see Figure 1.3), the hominids are clearly playing with a bigger deck.

Comprehending the human signature is no small task. Many different disciplines and points of view from both science and the humanities must contribute (Gazzaniga 2008; Greenspan and Shanker 2004; Buller 2005). Even then, singling out particular human traits as definitive always provokes arguments. So at the risk of being arbitrary, I shall simply list my preferences: cognition, prediction, self-awareness, syntactic language, bipedal locomotion, technology, aggression, altruism, culture, the arts, spirituality, religion, and science. You may prefer to add or delete items. If your list differs from mine, then your challenge, like mine, will be to show how each distinctive feature reflects, directly or indirectly, our species' evolutionary commitment to control.

Homo dominus and the Human Strategy

In this book, we look back from the twenty-first century to a distant rain forest several million years ago to witness the emergence of a new species, *Homo dominus*. Their challenge was to survive in a changing world. Their strategy was control. It rested on the molecular, genetic, and behavioral bedrock of life. It dwelt in the neural architecture of the brain and drove its behavioral outcomes. In its quantitative excess, this ancient control revolution was a qualitatively new phenomenon, the human strategy. It enabled our evolving ancestors to survive their changing world. To understand our origin as humans, we must understand the theory of control.

Keep in mind, however, that the first flickers of the human strategy probably started long before the emergence of recognizable hominids. Our earliest ancestors undoubtedly resembled the other great apes more than they did modern humans. What would have made them different, and perhaps more recognizable to us, was how they used their brains and behavior to deal with the world around them. As environmental instability took its toll on their peers, these incipient hominids thrived and multiplied. With each successful generation, their human genes accumulated and spread ever wider, conveying an increasingly differentiated set of characteristics that we human beings consider unique to ourselves. Our task in the remainder of this book is to work through these characteristics to understand how they made us what we are.

As we undertake this task, I remind you of the command issued to humankind by Linnaeus when he named us: *nosce te ipsum*. Know thyself. The act of questioning our origin is a reflection of the very strategy that catalyzed our origin: that we understand and accept who we really are and what we really do, so that we can minimize our negative consequences, and maximize our positive, as we grope our way forward as a species.

CHAPTER 2

Homo auguris, the Seer: Thought in a Changing World

I Think, Therefore I Survive

If I challenged you to name the one definitive trait that separates humans from the other apes, you would surely be tempted to say *thought* or *cognition* or some other variation on the general theme of an active intelligent mind. It would be hard to argue against this response as the breadth and power of human cognition appears vastly greater than the seemingly rudimentary forms exhibited by other species, at least insofar as we can measure them. If the power of thought were not a defining human trait, then we would have to attribute the same power to other species, but in a form unrecognizable to us. While sophisticated behavioral studies over the past three decades have produced considerable evidence for cognitive processes in chimpanzees, dolphins, and others (Premack and Premack 2003), it seems clear now that they are relatively limited, at best comparable to those of a bright human child. Of course, it is possible that we simply have not found the right instruments to assess cognitive capabilities in other species, nor have they chosen to reveal them. However, it seems more likely that other species simply cannot engage enough information processing capacity to build, manipulate, or display the equivalent of human thought.

At some point in the distant past, our ancestors on the margins of the great rain forest were probably equally limited in their cognitive capacities. Over the millennia, however, their cognitive capacities evolved until they clearly exceeded those of their deep forest cousins. Natural selection transformed these marginal apes into *thinking* apes. The question is, what was it about those early cognitive processes that enabled their carriers to do better than their less thoughtful contemporaries? How did our ancestors' existence become less precarious through the expedient of thought? Philosopher Kim Sterelny argues in his *Thought in a Hostile World* (2003) that the evolution of human cognition must reflect the selective advantages of comprehending the world differently and acting adaptively on that comprehension. This view rightly emphasizes adaptive action as the core substrate of natural selection. It is a species' ability to survive and reproduce that perpetuates its genes, not simply its capacity for thought. Nature did not reward our ancestors merely for thinking. It sustained them because their actions, based on the outcomes of those early cognitive processes, proved more adaptive than those of their less cognitive cousins. As a result, the genetic foundations of cognition proliferated, establishing the line of hominids that endures today. How cognition promotes adaptive behavior remains a fundamental question for cognitive theorists. The balance of this chapter offers some ideas.

In keeping with this book's convention of offering new names for the evolving human species, I shall refer to the first recognizably cognitive ape as *Homo auguris*, the Seer. This name reflects the distinctly human talent to see the world not only as it is, but also how it could be, how it should be, and sometimes even how it will be. The emergence of *seers* on the margins of the great Miocene rain forest marked a crucial step in human evolution.

Fundamentals of Cognition

Let's begin by clarifying some terms. Most researchers use the term *cognition* to refer generally to the act or process of knowing. The term usually implies the higher functions of the human mind, both on the perceptual side, as in concepts, categories, and abstractions, and on the action side, as in covert behaviors like planning, organizing, and imagining. The study of cognition falls within the purview of *cognitive science*, a loose association of scholarly disciplines that includes psychology, neuroscience, linguistics, anthropology, information sciences, and philosophy (Gardner 1985).

Historically, the study of human cognition followed two major paths. The traditional approach, which prevailed until the 1950s, held that cognition is primarily a human experience, that the process of thinking is inseparable from the experience of thought, that is, how it *feels* to think. Cognitive science has always struggled with this approach as it inevitably devolves to an introspective exercise of thinking about thinking. This makes for great philosophy but relatively poor science.

In the latter part of the twentieth century, a more rigorous scientific approach emerged when researchers reformulated cognition as a problem of information processing in complex systems. By manipulating inputs and measuring outputs, this new cognitive science sought to characterize mental processes as objective instances of information processing. By applying the rigor of mathematics, neuroscience, cybernetics, information theory, and neuropsychology to the internal workings of the human mind, modern cognitive science began to make progress toward understanding of its fundamental nature.

In this chapter and the following one, I shall address both approaches, since both ultimately contribute to our basic understanding of cognition as a human characteristic. While the modern information processing view has greater scientific value, it tends to ignore the experience of thinking, which probably every reader would agree is something truly remarkable. On the other hand, while the experiential view offers a more intimate link to the inner realms of the mind, it is less accessible to objective scientific analysis, and therefore less able to address the fundamental question of how cognition contributes to adaptive behavior. To understand the full scope of humanness, we need to understand both. So, in the present chapter, I shall address cognition in its operational information-processing sense. In Chapter 3, I shall overlay the experience of human consciousness and self-awareness on the underlying information processing system.

Operationally, we can define cognition simply as a complex set of information signals moving through assemblies of neurons in nervous systems, and interacting in various ways with an external environment. Our task is to build a coherent framework for analyzing this flow of information and to show how, more often than not, it leads to adaptive and reproducible behavior in real environments. My ultimate goal is to establish an objective model of cognition, that is, an explanation that rests entirely on material elements, observable, measurable, and manipulable in the real world, and on which natural selection might reasonably operate to transform a less cognitive precursor ape into a more cognitive human ancestor. Throughout this analysis, I shall assume that the natural environment of cognition is the human brain, and that its structures and connections underlie all the phenomena we associate with cognition. At the end of this chapter, I shall briefly address some of the unique characteristics of the human brain that neuroscientists believe are responsible for certain cognitive processes.

The above definition gives us three fundamental building blocks with which to assemble basic cognitive information processing systems: 1) an external environment; 2) overt behavior by individuals in that environment; and 3) internal information processing events that somehow connect the first two. Let's look at these elements more closely.

The contribution of the external environment to human cognition begins with biological elements selectively tuned to the energies and forces in the environment. For example, modern physics teaches us that all living things live in a vast sea of photons of varying energies. Most organisms have biological structures, such as photo-reactive proteins, that selectively interact with photons of certain energies according to the laws of molecular chemistry. The protein absorbs a photon, gains energy, and undergoes a change in conformation or reactive potential. Such interactions represent potential transductive events, that is, the transformation of external physical energy into internal biological information. Because these interactions can only take place in the presence of particular types of photons, their occurrence carries information about the distribution of particular types of photons in the external environment. For example, if the photons in question have wavelengths in the range of 400–650 nanometers, then their biological receptors may carry information about the current state of direct solar radiation, since that is a major source of photons at these visible wavelengths. Thus, an organism with such photo-reactive proteins has fairly reliable information on whether it is day or night. In effect, they *see*. In a similar fashion, over geologic time, the steady progression of such simple developments has produced a vast collection of biological structures and processes, genetically coded and physically rendered within the organism, whose interactions with the external environment can extract information about the environment. This capacity is the first essential step in the creation of adaptive information processing systems, such as human cognition.

Whether organisms can use the information they collect is the next critical question. It does an organism no good to *see* if it cannot *act* appropriately on what it sees. Thus, we have the second fundamental building block of cognitive evolution: adaptive action in the external environment. Action in this context refers to changes in the state or configuration of the external environment caused by the organism, or to changes in the organism's position or orientation relative to other elements of the environment. Thus, actions might include simple movements within and through the environment. Such movements might direct forces and energies to alter or relocate specific elements of the environment. Once executed, such actions may change the organism's exposure to other energies and forces in the environment from which it may extract new bits of information. Thus, action has the potential both to modify the environment and to change the individual's relationships with specific elements within the environment.

Almost all organisms have the ability to move within their habitats and to transform their environments in various ways. Such actions may be simple or complex. For example, a simple unicellular pond organism might have a rudimentary locomotor system, like a flagellum, that enables it to move from place to place in the pond. When the locomotor system is active, the organism can effectively counteract the pull of gravity and remain near the surface of the pond. When the locomotor system is inactive, the organism sinks to the bottom. Thus, the presence of an action system potentially alters the organism's ability to interact with its natural habitat. Unlike its nonmotile peers who must dwell at the bottom of the pond, the mobile species has options: it can live either at the bottom or near the surface. Mobility therefore increases the range of adaptive alternatives available to an organism. Starting with such simple mechanisms, the diversity of action patterns displayed by earth's creatures has grown vast and complex.

Taken together, the diversity of information gathering mechanisms and the diversity of actions represent an enormously large array of possible combinations. The crucial question for each such combination is whether it conveys adaptive value to the organism that exhibits it. Thus, in our pond example, if the increased opportunity to move to the surface increases the organism's exposure to light near the surface, and if more light translates into better reproductive prospects, then that action repertoire conveys a significant selective advantage. Less mobile organisms have less ability to compete, and eventually may go extinct.

If we now take the logical step of connecting action systems to perception systems, we have the crucial third element in the construction of cognitive processes: the *linkage* between information collected from the external environment and actions taken in that environment. Depending on the organism and the environment, such linkages may be simple or complex. In our pond organism, for example, the linkage may be no more that a light-sensitive protein binding directly to contractile proteins that drive the flagellum. When light is present, the binding conformation triggers movement of the flagellum. The organism may then move toward the surface to maximize its exposure to light. When light is absent, movement ceases, and the organism sinks to the bottom to conserve its energy. Here the *cognitive* linkage, if we can call it that, is simply a direct biochemical interaction between the perception and action systems. In more advanced species, the cognitive linkage may be much more complex, perhaps engaging many millions of elements. However, regardless of their relative complexities, both examples represent the same basic event in the information processing view of cognition: linking information in one part of the organism to action originating in another part.

Viewed as a sensory-motor linkage problem, we can immediately appreciate why cognitive processes seem inherently more complex than either perception or action alone. The linkage problem grows exponentially more difficult as organisms evolve greater sensory and motor repertoires. Suppose, in the pond example above, the successful organism evolves a new detector system for a particular chemical substance present in the pond environment, plus a modified second flagellum to produce, say, turning behavior. This more advanced organism now has two perceptual modalities (light and chemical) and two options for action (swimming and turning). In order to utilize all these capabilities fully, it must now have *four* cognitive linkages: the original (light-swimming) plus three new ones (light-turning, chemical-swimming, chemical-turning). Thus, the linkage problem has quadrupled, while the sensory and motor elements have each only doubled. This exponential relationship represents a fundamental evolutionary hurdle in the development of cognitive information processing systems. Each incremental addition of perception and action capacity generates an exponentially greater requirement for linkage capacity. Failure to keep pace in the linkage domain leaves an organism potentially vulnerable to biological competitors that do develop such linkages.

Now, much of the preceding discussion may seem oversimplified and irrelevant to the problem of human thought. However, it will help us understand what cognition is at its root level. It is, simply, an internal information processing system that connects two aspects of the external environment: one, the source data streaming in from all parts of the environment, and two, the transformation of those data streams by the organism's own presence and actions in the environment. To the degree that such connections produce reproducible adaptive outcomes for a given species, they represent evolutionarily significant advantages. What distinguishes one species from another in the cognitive domain is simply the relative magnitude of the three basic cognitive building blocks: sensory diversity, range of action, and the internal linkages between the two.

Of the three, the linkage component is clearly the most relevant to the problem of human cognition. If we compare human capabilities with those of our nearest biological relatives, the chimpanzees, we find that in the sensory domain, chimpanzees are comparable to humans in all basic modalities: vision, hearing, touch, taste, and smell. In the action domain, the differences are somewhat greater, but still within a reasonable range of variation. Chimpanzees are typically stronger and quicker than the average human, but they lack the human's fine motor skills, particularly in the hands and glossopharyngeal areas. The major difference between the two species lies in the vastly greater quantity and quality of linkage elements in human cognitive system. From the same environmental data, the human's cognitive system can seemingly extract and utilize greater amounts of information than the chimpanzee's, and it can connect that information to its action repertoire with a greater likelihood of a favorable outcome. While chimpanzees represent quite a capable species in other areas of life, they fall far short of humans on the cognitive linkage dimension.

We come then to the basic understanding that the emergence of human cognition, as displayed in our theoretical construct *Homo auguris*, lies in an enhanced capacity for linking information that originates in the external world with action in that same world. These enhancements derive from physical and chemical changes within the primary organ for processing information, the brain. How and why an ape species might have moved from its ancestral apelike complement of linkage structures to a more elaborate, and biologically more expensive, set of human cognitive linkages remains our basic question. The answer lies, in part, in the organization of cognitive linkages in the evolving brain. The next section addresses this issue.

Neurophilosophical Linkages

There are many ways to connect perceptual and action systems in living brains. Over the past five decades, cognitive scientists have proposed a great many hypotheses on how the human brain actually accomplishes this. However, consensus remains elusive. Indeed, as vast and complex as cognitive science has become, it may actually need to get *more* vast and complex, as Marvin Minsky (2006) recently suggested, because attempts to explain the workings of the mind in terms of simple one-size-fits-all principles have yet to succeed. A proper review of all the proposed models of human cognition, in detail, is far beyond the scope of this book.

For my purposes, a better approach is to focus on some broader concepts of cognitive organization, in the hope of narrowing the problem to a few select principles that underlie human cognition as a product of evolutionary forces acting on an ancient population of ancestral apes. Recall that before there was full-blown human cognition, there must have been more rudimentary apelike forms that conveyed some selective advantage to their bearers over their relatively less cognitive peers. Understanding how the modern human form might have emerged from such prototypical apelike forms may help us focus on the truly relevant principles of cognitive organization, leaving the esoteric details for another time.

Let's start with one of the classic models of cognitive organization, as depicted in Figure 2.1. This model suggests that the substrate of cognition lies between the perception and action systems, forming an obligatory third layer. This is the so-called *classical sandwich* view. Here the cognitive layer presumably takes input from the perceptual apparatus and delivers output to the action apparatus, but the internal processes within that cognitive layer

33

remain functionally separate and largely independent of the other two layers. True cognitive science, according to the sandwich model, focuses on the inner workings of this middle layer. The perceptual and action layers are merely input/output services that attend the cognitive engine, but do not themselves represent true cognitive processes.

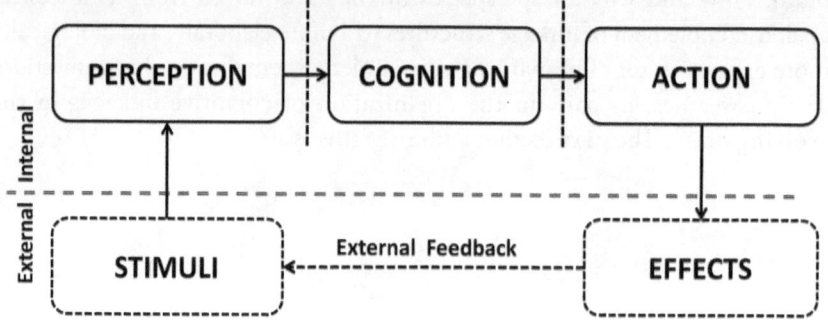

Figure 2.1. The *classical sandwich* model of cognition. The traditional view is that cognition resides in a layer of neural functionality sandwiched between perception, from which it takes input, and action, to which it sends commands. The processing arrows are strictly one-way as depicted. This model contrasts with the view of cognition as a function of dynamic interactions among these same components.

While many cognitive scientists might agree with this depiction, it has recently come under criticism by theorists working in the realm of neurophilosophy. This is a relatively new academic discipline that addresses fundamental philosophical problems, both classical and modern, in the context of neuroscientific data (Churchland 1986). Neurophilosopher Susan Hurley (2001) has systematically exposed the flaws in the classical sandwich theory. Hurley contends that this traditional theory drifts off course when it tries to separate cognitive processes from perceptual inputs, that is, information from the external world streaming in through the body's sensory apparatus, and from action outputs, that is, information flowing out to the external world through the body's effector apparatus. While the classical sandwich may appeal to many cognitive scientists, Hurley musters considerable independent data and analysis to suggest that the phenomena we classify as cognitive are in fact inseparable from the dynamic interactions of perception and action. Indeed, cognitive functions are actually emergent properties of the *interplay* between perception and action, each coequal and inseparable from the other. Thus, the supposed filling of the classical sandwich, and therefore the supposed independent substrate of cognition, does not exist, according to Hurley's analysis.

Hurley's work reinforces that of many others (Churchland 1986; Jackendoff 1987; Hutchins 1995; Proust 1999; Bickhard 1993, 1999) suggesting that cognition is actually a manifestation of the interactions between perception and action systems, that they are integral to cognitive processing and inseparable from it. By extension, Hurley suggests that even perception and action, once linked together in an interactive cognitive system, no longer operate independently. As perception and action become inextricably intermingled, each must assimilate properties of the other. That is, brain systems to which we attribute perceptual functions are, to some degree, already engaged in action repertoires. Similarly, brain systems to which we attribute action functions already carry perceptual information and influence perceptual processing. Only a relatively few sensory neurons flowing in from the distant peripheral tissues legitimately carry the label *sensory-only*, and only a few motor neurons flowing out to the muscles are *motor-only*. Everything else, including the bulk of the nervous system, is a tangle of dynamic interconnections in which the participating elements must necessarily lose any pure and independent functional identity. Floating over this turbulent sea of interactions are the clouds we recognize as perception, action, and cognition, but for which we can find no unambiguous physical substrate within the interactive mass.

The neurophilosophical view does not deny perception, action, or cognition as definable entities. Information about the external world certainly exists in working nervous systems (perception), as does the impetus to move or act in or on the external world (action). That these two elements interact in various ways (cognition) is also indisputable. What is not so clear, according to Hurley and the other neurophilosophers, is that these functions have neatly compartmentalized physical substrates whose interrelationships can be traced simply by connecting the dots. Rather, the proper model for human cognition comes from the class of dynamic interactive systems. This view is reminiscent of recent conclusions from neuroscientific research on the supposed hierarchical organization of the nervous system, as outlined in the next section.

Hierarchies and Their Problems

Like the classical sandwich model, one of the longstanding views of human cognition is that its basic architecture is *hierarchical*. The hierarchical concept invokes the reductionist view that complex perceptions, actions, and thoughts arise through the progressive convergence of simpler elements. Upper levels of a hierarchy imply greater convergence than lower. Each step up represents an incremental enrichment of information content and cognitive abstraction. Only the number of steps limits the maximum complexity attainable in such a system. Cognitive theory traditionally asserts that the highest functions of the human mind emerge at the highest levels of the hierarchy.

The hierarchical view has strengths and weaknesses. On the one hand, decades of anatomical and neurophysiological study demonstrate clearly that nervous systems display the kind of functional convergence necessary to create higher levels of complexity from simpler elements (Gazzaniga, Ivry, and Mangun 1998). On the other hand, recent research suggests that the nervous system shows a much stronger tendency toward reciprocal connectivity and interactive loops than previously believed. Loops complicate hierarchies. Thus, while simple stair-step hierarchical processing remains an enduring theoretical construct in cognitive theory, we must be cautious in interpreting cognition strictly in these terms. The following sections summarize some traditional ideas about cognitive hierarchies and their problems.

Perceptual Hierarchies. Viewed from the periphery of the nervous system, the hierarchical model appears unassailable. The body's sensory systems comprise an assortment of individual sensors and transducers scattered throughout the skin, muscles, joints, inner ear, tongue, nasal passages, and the retina of the eye. Each has but one function: to turn energy or matter in the local environment into bits of information that flow into the nervous system. Like automatons, sensory transducers repeat the same conversion over and over, operating in seeming isolation from other transducers, and from the higher centers of the brain. Their signals, multiplied many fold, produce constant streams of information flowing into the brain, representing the raw bit maps of the external world.

Arriving in the central nervous system, these bit streams undergo hierarchical convergence and transformation into complex perceptions. The pioneering work of David Hubel and Torsten Weisel in the visual system illustrates the process. These researchers inserted ultra-thin electrodes into the brains of experimental animals and recorded the activity of individual neurons in various visual areas (Hubel and Wiesel 1959, 1962, 1968). They found that visual processing follows a multi-step path. At each step, the activity of individual visual neurons seemingly reflects the progressive convergence of the information created in prior steps. Thus, neurons lower in the convergence are little more than simple bit map carriers, while those higher up respond to features in the environment, like edges, lines, or polygons. Further convergence produces more complex detectors that encode movement, color, stereopsis (differential input from the two eyes), subjective shape contours, and other complicated patterns. Moving up still further, visual information combines with information from other sensory modalities to create multimodal perceptual maps.

This traditional view of input convergence remains a rich area of scientific inquiry. Neurophysiologists continue to trace pathways and connections, giving us increasing detail about the transformation of incoming information. However, cognitive theorists have always wrestled with the vexing question of when, where, and how such sensory processing turns into perception. For example, if we observe a bird flitting through the trees, our peripheral sensory systems deliver steady streams of packetized information about the event. Visual data flows from the retina, auditory signals from the inner ear, perhaps even tactile information from the skin as the bird fans the air. If we inserted electrodes into our brains, we would measure these isolated bits of information about the bird, perhaps correlated with each other, but clearly separate and limited. The question is whether there are individual neurons somewhere in the brain that represent the whole bird in flight. And if so, what do these *flying bird* neurons converge on? Do we have still higher convergence to, say, *birds flying between trees* as opposed to *birds flying south for the winter* and so on? At what point does this presumed perceptual hierarchy break down? And what cognitive entity does it become at that point? Flying birds certainly exist in the real world, and their presence clearly triggers our sensory transducers to send signals into our brains, but we have yet to find the places in the cognitive hierarchy where our perceptions of whole birds in their natural environments reside. This has always been a problem for the theory of perceptual hierarchies and remains one today.

Action Hierarchies. Mirroring the perceptual system, but reversed in direction, is the action, or motor, system. This system also reflects hierarchical information processing, but the transducers in this hierarchy lie at the *end* of the signaling chain rather than the beginning. This complicates the logic a little. Where the sensory system sends bit-mapped information into the central nervous system to become convergent information, the motor system must reverse this flow and parse highly convergent information into bit maps of the organism's peripheral musculature.

Muscles are the body's action transducers. There are two major types: (1) skeletal muscles, which attach to the bones and mediate the individual's overt movements; and (2) smooth muscles, which attach to the body's internal organs and mediate various autonomic functions. Muscles comprise bundles of contractile filaments that have one simple function: to contract in response to neural signals. The body's muscles receive signals from the brain, spinal cord, and autonomic ganglia via the peripheral motor nerves, the final common pathways of action. Signals in the brain reach the muscles by flowing down a presumed action hierarchy, out through the peripheral nerves, and into the muscles. Each muscle fiber receives only a few motor neurons. The arriving signals reflect a highly distilled extract of the entire brain's recent information processing activity, including the highest functions of the mind, all boiled down to specific movements of specific muscles at specific moments in time. While peripheral motor systems mediate all movements, they are only the robotic executors of action patterns defined, assembled, and coordinated at more remote parts of the system.

We know a great deal about the lower levels of action processing. Groups of spinal motor neurons send connections out to form synapse-like junctions with muscle fibers. When activated, each neuron causes its target muscle fiber to contract. When these neurons fire as a group, the entire muscle contracts, which moves the limb. Comparable neurons innervating opposing muscle groups operate in a similar fashion to move the limb back. At the same time, sensory neurons embedded in the muscles and joints, detect changes in the position and tension of the muscle and provide feedback signals to spinal motor cells to help control and smooth the movement. Replicated many thousands of times across virtually all muscle groups, these reflexive control loops provide the foundation of the skeleto-muscular movement system.

Moving backwards up the hierarchy, the spinal motor centers receive inputs from neighboring spinal neurons and from various centers of the brainstem. Some of these interactions provide timing and coordination for highly programmed movements such as locomotion. Other centers in the brainstem and basal ganglia superimpose additional coordinating and integrating functions to effect more diversified sequences of movements.

Moving still further back up the action hierarchy, the cerebral cortex overlies the lower movement systems. Several cortical areas are involved, including the *primary* motor area, *premotor* and *supplementary* motor areas, and the massive *prefrontal* cortex. These areas appear to be less involved in detailed movement programs, and more in the selection, preparation, and initiation of movement patterns appropriate to particular dynamic environmental situations. The prefrontal cortex in particular is an area of crucial importance in the study of human cognitive functions. Decades of research have suggested that this area has the requisite properties and connections as the primary center for cognitive linkage functions (Warren and Akert 1964; Miller 1999; Deacon 1997; Frith and Dolan 1996; Gabrieli, Poldrack, and Desmond 1998). Some theorists consider the prefrontal area an executive decision center for voluntary behavior in response to perception (Goldberg 2001).

Cognitive science has traditionally paid less attention to the action side of the functional brain, preferring the perceptual side. Somehow, the mental imagery associated with perception always seemed to fit our notions of cognition better than actions and movements. That bias now seems to be falling away, as cognitive theorists increasingly see action as an integral component of cognition (Hurley 1998, 2001, 2006). Indeed, the word *think* is a verb, suggesting mental action, as in "I am going to think about this tonight." Similarly, cognitive concepts like *visualizing, imagining, dreaming, planning, preparing,* and so on, all carry the connotation of action, albeit without overt movement. The emerging view is that mental actions are as important as mental perceptions in cognitive processing. Each plays an essential role: perceptions *construct* the world in our heads, while actions *deconstruct* it into adaptive outcomes.

Problems with Hierarchies. While simple stair-step hierarchies have some appeal in traditional cognitive theory, they have major flaws when applied to real nervous systems. Two significant problems are 1) *reciprocal connectivity* between presumed hierarchical levels; and 2) lateral connectivity between adjacent hierarchical modalities.

The problem of reciprocal connectivity is amply illustrated in the visual system, perhaps the best-documented area of sensory processing. Extensive neuroanatomical and neurophysiological studies (Felleman and Van Essen 1991) have shown that visual centers in the brain are *reciprocally* interconnected. Each center both sends and receives connections from the other centers in the system. Thus, far from being a simple stair-step hierarchy, the visual system appears to be much more dynamically organized. For example, in the primary visual area (V1) of the cerebral cortex, only about twenty percent of the incoming connections come from brainstem visual centers, logically the next lower layer in the hierarchy. The remainder comes primarily from neighboring cortical areas, including the next higher visual level (V2), which appears to send as many connections *down* to V1 as V1 sends *up* to V2. In other words, the flow of visual information, even at these relatively low levels of the presumed hierarchy, is seemingly not hierarchical at all. Sillito and Jones (2002) have shown that information flowing *down* the presumed visual hierarchy actively influences sensory information processing at the earliest subcortical stages. Thus, neural areas considered to be higher in the processing scheme apparently send substantial volumes of information to supposedly lower centers. Clearly, the simple stair-step structure we might have envisioned earlier does not capture what is really happening in the visual system. Ongoing research suggests that reciprocal connectivity may be the rule rather than the exception in neural organizations.

Lateral connectivity is also a problem for hierarchical models. This is a relatively new area of neuroscientific study, rich in implications for theories of cognition (Macaluso 2006; Schroeder and Foxe 2005; Bell et al. 2005; Calvert, Spence, and Stein 2004; Mesulam, 1998). Lateral connectivity refers to the property of neural systems to send and receive lateral connections to and from other processing systems outside their presumed hierarchies. While it has always been assumed that the neural processing of different sensory modalities (vision, hearing, etc.) eventually converged at the higher levels of the cognitive hierarchy, it is now becoming clear that cross-modal integration occurs at much lower levels of sensory processing. This means that from the earliest stages, the different perceptual modalities are already exchanging information with each other. Moreover, these cross-modal exchanges have significant impacts both on how perceptual information in processed within any given modality, and on how that information generates behavioral outcomes. Thus, almost from the first connection, the notion of independent sensory modalities arrayed in their own separate hierarchies appears to be incorrect.

The implications of ubiquitous lateral connectivity are important. It means that our notions of, say, *pure* visual information traveling along one dedicated neural hierarchy, and *pure* auditory information traveling along another, is inaccurate. The visual and auditory systems appear to have physical interconnections at very early stages in their respective sensory processing chains. This means that the perceptions they mediate must interact almost from the beginning. They are no longer pure sensory channels. Visual processing cannot be exclusively about *seeing* if it also carries embedded auditory or somatosensory signals. Conversely, auditory processing cannot be exclusively about *hearing*, or somatosensory processing exclusively about *touching*, if each also carries embedded visual signals. If lateral connectivity is part of their underlying structures, then all perceptions in all modalities must carry elements of all other perceptions in all other modalities. If the same patterns apply on the action side of the nervous system, then the intermingling of perception and action, which is the cornerstone of cognition, is a salient feature of the nervous system almost in its entirety. The supposed *lower* centers of brain are therefore no less involved in cognitive processing than the supposed *higher* ones.

Reciprocal and lateral connectivity, now clearly established experimentally, complicate any simple notion of information hierarchies. Although the convergent build-up of complexity remains a compelling model of neural processing, its actual mechanisms appear more complicated and dynamic than originally thought. If information at any given point may flow in all directions—up, down, and sideways–it is almost meaningless to assign hierarchical descriptors like *lower* or *higher*. The reciprocity and laterality exhibited in real brain architectures suggests that circularity may predominate over linearity. While functional hierarchies may still exist, they appear to be embedded in, or emergent from, a matrix of interacting loops. Simple stair-step interpretations must therefore give way to dynamic nonlinear models.

Fortunately, there is a substantial body of theoretical and experimental work that addresses such problems. It is control theory.

Cognition and Control

As problems emerge for the traditional views of human cognitive organization, as embodied in *sandwiches* and *hierarchies*, researchers and theoreticians are now looking at more dynamic models, such as those implied by control theory. The control schematics presented in Chapter 1 clearly define an information processing system that includes the three basic building blocks of cognition: 1) the transduction of forces and energies from the external world into information; 2) the activation of effector systems that transduce information back into external forces and energies; and 3) the linkage of information and action through various convergence, integration, and transformation processes. Such control architectures therefore represent, by definition, cognitive information processing systems. We must therefore include them as candidate solutions to the problem of human cognitive organization.

Obviously, however, the more interesting question is not whether control systems meet the minimum definition of cognitive processing, but whether they are actually *synonymous*. Does human cognition actually utilize neural control architectures in the human brain? The idea has some support. Most notable is again Susan Hurley (2006), who has assembled an impressive review of neuroscientific evidence to support a five-layer cognitive model, called the "Shared Circuits Hypothesis." Hurley's theory incorporates basic and enhanced control architectures similar to those outlined in Chapter 1, plus some additional features. Her analysis supports the basic thesis that cognitive organizations can indeed build on well-known control principles and architectures. The breadth of experimental data assembled by Hurley to support her theory suggests that control interpretations may actually offer the best available approach to the study of human cognition.

Hurley's analysis builds on earlier work on applying basic control principles, such as those illustrated in Figure 1.1, to the phenomena and organization of human cognition. Control theorists realized early on that modeling real-world human behavior required greater complexity than simple feedback control loops. One had to account for the appearance of hierarchical organizations, as well as the emergence of supposedly higher mental faculties such as imagination, prediction, and planning. Simple feedback control, while almost certainly participating in such cognitive phenomena, falls well short of providing a satisfactory explanation. As a result, several behavioral theorists, notably William Powers, proposed complex control architectures to address complex behavior (Powers 1973, 2008; Marken 1986; Cziko 1995, 2000). Figure 2.2 illustrates this concept using in a stacked interactive array of simple control modules. Let's examine this structure more closely.

First, we note that the inputs and outputs of this array each form a kind of hierarchical stack. On the input side, each functional level adds a layer of sensory convergence, like the Hubel-Wiesel model of the visual system. Similarly, on the action side, information cascades downward toward the final motor links to execute movements in the external world. Of course, given our above discussion of hierarchies, we should take care not to interpret this hierarchical appearance too literally. The flow arrows are therefore not uniformly upward on the perception side nor are they uniformly downward on the action side.

Linking these perception and action arrays is a set of control modules. Each of these modules operates in the same way as the one depicted previously in Figure 1.1. In any given module at any level of the array, perceptual (P) signals enter and action (A) signals exit. Within the module, the P signals combine with reference expectations in a comparison operation, from which a dissonance signal emerges. This, in turn, may trigger an output, or A signal. As this signal exits, it sends a signal back into the input processing system, creating a closed-loop feedback system embedded within the larger array. These same basic operations occur in all control modules in the array.

Now, imagine all the various classes of potential interconnections among these elements, including direct, reciprocal, and lateral. With that in mind, you should be able to convince yourself that, even though this stacked control architecture has the appearance of a hierarchy, it can actually function as a relatively flat information-processing array. From any given point, you should be able to trace a pathway to virtually any other point. Points supposedly *higher* in the array can send information to lower points. Points on the left, or *perceptual*, side can connect to those on the right, or *action* side and *vice versa*. In some case connections occur directly within the local functional domain, while in others, they happen through intermediary pathways, such as dissonance or feedback connections. This exercise illustrates that a single system, depending on the relative dynamic strengths of its signaling elements, can exhibit attributes either of a hierarchical organization, or of a relatively flat functional array. Control architectures, as depicted here, have dynamic, nonlinear properties that enable much greater flexibility and functional diversity than simple stair-step models.

Figure 2.2 also illustrates William Powers' hypothesis about the origin and nature of reference expectations in complex control architectures. The figure depicts expectations arising from three sources: 1) instinctive processes, such as hunger or thirst; 2) previously learned events or sequences; and 3) action/output signals streaming from one control module to another. This last source reflects Powers' original suggestion that the outputs at higher levels of the control array may define reference expectations for lower levels.

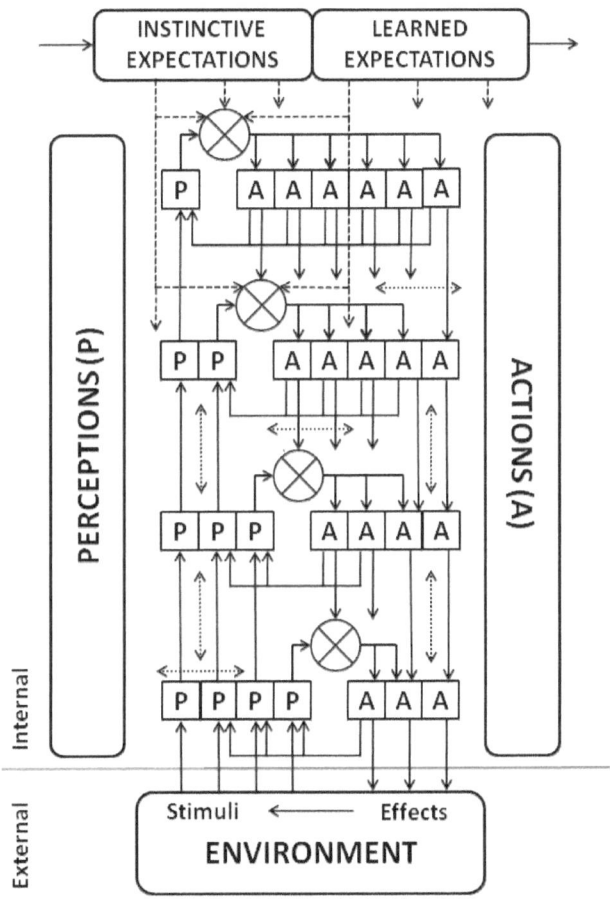

Figure 2.2. A stacked control architecture reconciles apparent hierarchical processing with the dynamic interaction model of cognition. Basic control modules are ubiquitous in the nervous system, some at supposed *higher* levels and some at *lower*. The lateral interactions between perception and action through dissonance and feedback pathways, combined with reciprocal connectivity among virtually all cortical areas, can transform such apparent hierarchies into flatter functional arrays.

Thus, while each module still processes information in the same way, the specific comparison operation in any one module may sometimes involve the direct convergence of perceptual inputs with action outputs. In other words, the impetus for action at one level may actually determine the expectation of perception at another level. Dissonance generated from these comparisons may in turn trigger new action outputs, which in turn set new reference states for the subsequent levels, and so on down the line. In Powers' original view, this multi-layered cascade theoretically extends all the way to the final common pathways for muscle movements and overt external actions. As these actions occur, their effects propagate through the environment, producing new waves of perceptual processing sweeping back up the stack. At every point, these new perceptions compare again with local reference signals and new dissonance patterns emerge, which in turn send new waves of reference activity back down the stack, and the cycle repeats. The system continues to send perceptual waves up and reference/action waves down the control stack as the individual negotiates a complex world.

Powers' concept of how dynamic expectations might form offers a unique interpretation of how cognitive linkages determine behavior. Instead of issuing commands at the top of some presumed motor hierarchy, cognitive outputs may simply set expectations for perceptual inputs elsewhere in the system. As Gary Cziko (1995) put it, such cognitive processes do not tell us what to *do*, but what to *perceive*.

The fact that expectations can arise from multiple sources at multiple levels in the brain is crucial to our understanding of human cognitive evolution. From basic biological drives to the potentially huge volume of learned expectations stored in a single brain, the proliferation of this fundamental component of control systems defines the primary thrust of human cognitive evolution. Any event in an individual's life, from trivial to profound, has the potential to serve as an expectation in some future situation. It depends only on how the brain stores the memory trace, and how it subsequently interacts with other neural elements. If it remains active and functions in the brain's control architecture as an expectation, then it must influence behavior as we have described above. Learned expectations, as we shall see in later chapters, represent the major source of differentiation between human behavior and that of the other ape species.

Although control architectures have intriguing features that clearly bear on our concepts of human cognition, it will take much further research to validate such models as the primary basis of human cognition. Until that happens, traditional cognitive theorists may still cling to the view that the *higher* functions of the human mind remain separate and special. Similarly, traditional behaviorists may continue to emphasize objective actions as the only meaningful source of data on the mind. In this book, we shall take the view that both are valid and essential. Cognition and behavior are manifestations of the same underlying systems that have evolved to produce adaptive solutions to complex problems. Neither abstract thought nor elemental behavior has preeminence in the evolution of humanness.

The Emergence of the Seers

To understand the evolutionary origins of human cognition, we need to specify what precisely makes it so valuable that a species would devote so much biological differentiation to achieve it. We presume that the earliest forms of *Homo auguris* held some crucial selective advantage over their less cognitive peers, but what precisely did they do differently? Scientists, sociologists, and philosophers have debated this question for decades and the list of candidates is long (Sterelny 2003).

Like many others (Llinas 2001; Hawkins and Blakeslee 2004), I believe that *prediction* needs to be near the top of any such list. The notion that individuals can somehow look ahead in time, and adjust their actions according to what the see, seems truly remarkable. Indeed, it would be unbelievable to us if it were not something we do routinely on a daily basis. How did such a remarkable ability arise?

Prediction in the present context means the general ability to extract and retain information about a given object or situation, and to utilize such information beyond the immediate spatial and temporal constraints of the situation. Prediction implies a greater depth of perception, a more flexible and transportable comprehension of causality, and an ability to project the outcomes of action in advance of action. In general, actions based on prediction should have advantages over reactions to unpredictable stimuli. The error rate in the former should be lower than in the latter. If errors represent fatalities or reproductive failures, then a species whose brain mechanisms include a predictive function, however rudimentary, should succeed more often than those without such a capability. As predictive success translates into reproductive fitness, its underlying genetic substrates of should endure and expand.

The ability to predict events and take action to improve their outcomes is clearly a cognitive function of seemingly high complexity. However, if cognition is a consequence of control processes, as suggested above, then we need to look for predictive capabilities in the functional architecture of control.

One possible mechanism is shown in Figure 2.3. This schematic shows the same basic control module you have seen before. However, I have highlighted an important feature not previously noted: specifically, that there are always *two* feedback paths in such systems. The first, *environmental*, involves information generated by an individual's actions in the external environment. This is the primary path in the original formulation of perceptual control theory (Powers, 1973). Presented with a particular set of external stimuli, the individual's control architecture executes its comparison operations and generates action in the external world. These actions have physical consequences that alter the pattern of external stimuli, thus closing the control loop through the external environment.

The other feedback pathway is *internal*, as shown in Figure 2.3 as a direct link from the action system into the perceptual system. Powers called this type of feedback the "imagination connection." Operationally, this

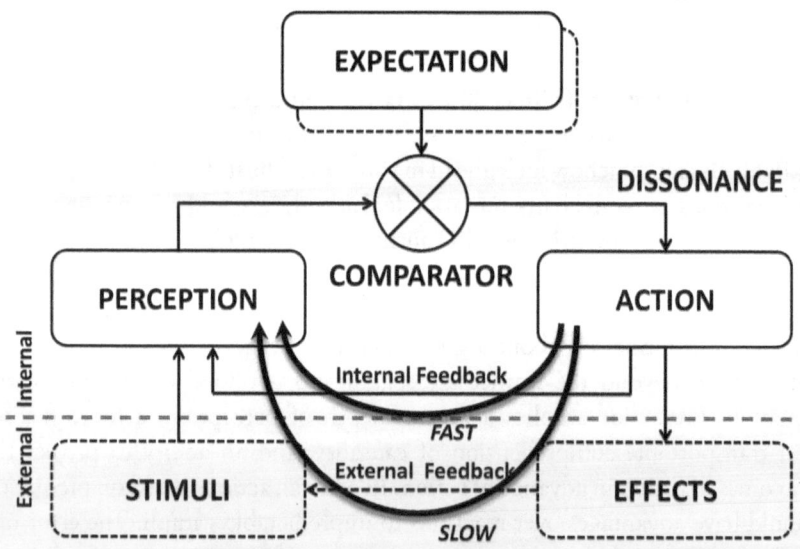

Figure 2.3. Predictive functions in multi-path feedback control systems may arise from the differential latencies of the respective feedback signals. Here the feedback information traveling on the internal path alters perception much sooner than the information coming through the external route. If the two signals are correlated, that is, they have consistent or reinforcing effects on perceptual processing, it is theoretically possible for the faster signal to act as a *predictor* for the slower one. In the complex, multi-path control architecture of the human nervous system, differential feedback latencies may provide the primary predictive mechanism that promotes successful anticipatory behavior.

internal feedback loop provides a direct route for action signals to modulate incoming perceptual signals, without necessarily going through the external environment. In effect, this internal feedback connection bypasses the external world to provide direct information to the perceptual apparatus about ongoing or upcoming actions.

What is important about these two feedback pathways is their potential time disparity. The external pathway requires actual overt movements in the external environment. The timeframe of such changes may vary from a few seconds to days, months, even years. The internal feedback path, in contrast, exerts its effects through neural signals traveling within the nervous system on a timescale of perhaps a few hundred milliseconds. Thus, the two feedback pathways have distinctly different temporal dynamics. This disparity allows the possibility of a predictive function for one path relative to the other. Both feedback loops start at the same point, the initiation of action, and both ultimately converge on the same point, the processing of sensory information. However, the signals in the internal path typically arrive sooner than those in the external path. From the perspective of that convergence point then, signals from the faster path could theoretically predict the eventual arrival of those from the slower. In other words, control systems with at least two feedback pathways have a built-in predictive potential, assuming some difference in the relative latencies of the feedback signals. Since both pathways could theoretically carry the same basic information (i.e., the consequences of action), the faster path could potentially serve as a predictor of what is coming in the slower.

Predictive signaling of this type is quite familiar to neurophysiologists. For example, the eye movement centers in the brain utilize collateral feedback to modulate visual inputs, a phenomenon known as an *efference copy*. The oculomotor system, which moves the eyeballs, sends advance *copies* of its motor signals directly into the visual system to neutralize the perceptual disruption of the impending eye movements. These signals arrive before the motor neurons produce the actual eye movement. From the perspective of the visual system, then, this efferent feedback signal effectively *predicts* the upcoming shift in the visual fields. These neutralizing signals have the beneficial effect of maintaining perceptual constancy during eye movements. When we move our eyes, we do not perceive the world jumping from place to place, even though the projection of the world on our retinas does in fact jump from place to place. Cognitively, we perceive the world as static while our eyes move, not the other way around. If our perceptual control systems did not make this adjustment, we would experience every eye movement as a sudden, dizzying shift in the world around us. Like the rider on a train who perceives the station moving while he sits still, an individual without such feedback compensation would be unable to distinguish the movement of external objects from their own eye movements.

Efference copy mechanisms probably exist at many levels of the nervous system, and may provide a useful model for our predictive propensities at the cognitive level (Grush 2003; Johnson 2000; Wexler, Kosslyn, and Berthoz 1998; Kosslyn and Sussman 1995). Predictive feedback signaling is also an essential feature of Susan Hurley's shared circuits model of human cognition (Hurley 2006).

Applying this same logic to the multilevel control array in Figure 2.2, we can generalize such predictive relationships to any number of levels where feedback signals originate and terminate at the same respective points, but arrive via different temporal and spatial paths. If we rank order these multiple feedback paths by their time delays, we could theoretically generate a multilevel predictive hierarchy, with the higher (faster) levels carrying predictive information about lower (slower). The role of such ubiquitous predictive feedback is a major feature of other models of cognitive functioning (Hawkins and Blakeslee 2004).

If predictive signaling is a natural property of multi-feedback control systems, and multi-feedback control systems represent the foundation of cognition, then we may hypothesize that the first cognitive ape, *Homo auguris*, emerged with a rudimentary capacity for prediction already embedded in its growing cognitive architecture. With the benefit of incrementally more information about the future consequences of action, these apes would have been able to function in complex environments with greater efficiency and fewer errors than their peers. Such incremental gains, however small, would have conveyed incremental improvements in their decision-making. While physically these early hominids might have looked much like their ape cousins, a keen observer might have detected those subtle differences in how they behaved. Perhaps they demonstrated a better anticipation of dangers and opportunities, as if they saw more clearly the result of action in advance of action. As these incremental successes gave them a better chance to survive and reproduce, their predictive functions endured and grew, bringing steadily more of the world's complexity under their growing predictive view. Eventually, their behavior became recognizably different from that of the other apes, and they emerged as *Homo auguris*, the Seer.

The Correlation Problem. There is a catch, of course. Prediction has value only in direct proportion to its accuracy. In our schematic in Figure 2.3, the internal feedback signal can indeed alter perception, but it is not clear how such changes could actually correlate with the real perceptual signals that will eventually come in from the external environment. If the internal predictive information does not correlate with the soon-to-arrive external information, then it is at best meaningless noise and at worst a disruptive hallucination. Imagine, for example, what would happen if our efference copy mechanisms for eye movements went awry and began injecting hallucinatory signals into our visual systems. Every time we moved our eyes, a wave of bizarre and uncorrelated visual signals would flood our perceptions, leaving us effectively blind to real external inputs. This hardly seems an adaptive outcome.

The crucial question then is how the system can establish a correlation between the two feedback paths in Figure 2.3 so that the faster path serves as a reliable predictor for the slower one. Two important areas of neuroscientific research offer potential answers: 1) pattern formation during childhood brain development, and 2) adult learning and memory.

Childhood Development. Neuroscientists have long recognized the effects of early experience and environment on brain development (Diamond 1988). Cognitive control systems do not arise exclusively through genetic hard-wiring. Integrating actions and perceptions appears to require a lengthy period of trial and error in childhood. Developing children start with only minimal control capabilities, displaying the characteristic clumsiness and bewilderment at how things work. Over time, however, they develop the smooth, integrated control of movement and perception characteristic of the mature adult. Experience during childhood appears to guide and shape perceptions and to build coordinated actions to deal with them. Human beings undergo an especially long infancy and childhood, perhaps reflecting the especially large number of connections and correlates needed for our highly developed cognitive and predictive capacities.

Neuroscientists suggest that the developing human brain is modular (Gardner 1983; Fodor 1983; Mithen 1996; Pinker 1997; Premack and Premack 2003), although some questions remain (Buller 2005). The basic notion is that the infant human brain starts with various prewired functional templates that respond selectively to particular patterns of stimulation and types of information. Such modules reveal themselves in children's innate expectations about certain stimuli, and their predisposition to build behavioral responses appropriate to such stimuli. Early experience with objects and events translates smoothly and rapidly into adaptive behavior appropriate to the relevant stimuli. Most theorists believe such modules reflect the genetic foundations on which adult behavior builds.

Such modular theories of child development emphasize the early interaction of perception and action. Successful development of adult capabilities requires a child's active immersion in their perceptual environment. To achieve competency, a child must repeatedly initiate actions and perceive their consequences in various situations. Frequent trials generate both internal and external feedback to the developing perceptual apparatus, eventually producing a network of perceptual correlates related to the consequences of actions in the natural environment. For example, a child who repeatedly drops objects and perceives them falling eventually develops an internal correlate between the action of releasing an object and the set of perceptual outcomes associated with falling objects. With repetition and development, the internal feedback associated with the action of releasing the object (e.g., activating the hand's extensor muscles) eventually trigger a virtual representation in the perceptual system of the object falling, even *before* the extensors fire and the object falls. In effect, the child begins to *see* the object falling before it actually falls. Mechanistically, this happens because the child's internal feedback pathways have developed a *predictive correlation* of the object falling. This virtual perception represents new input to the child's control system, which may trigger alternative actions, such as gripping the object more tightly or shifting its position. With this new action repertoire, the object stays safely in the child's grasp. Without it, the initial impetus to release the object goes uncorrected, and the object falls.

As children mature, their experiences with objects, situations, and events becomes progressively more complex. The associated build-up of correlative information between actions and their effects on perception continues. The greater the set of correlated experiences, the more likely the predictive success. Thus, children raised in highly enriched environments with greater stimulation develop stronger predictive correlations than children raised in impoverished environments. Lacking sufficient early experience, such deprived children may suffer from diminished anticipation and deficiencies in prediction, as if they are unable to suppress inappropriate actions or comprehend their consequences.

Learned Associations. Once the juvenile nervous system reaches maturity, its predictive correlation strategy changes. Rather than continuing to install hard-wired patterns into modular neural structures, the nervous system shifts to a more flexible process based on adult learning and memory.

Put in an entirely new environment, with completely novel physical and cognitive challenges, most adults act like toddlers, displaying the same lack of control and coordination, and often the same bewilderment at how things do or do not work. The same cause applies: in a totally new environment, the predictive correlations between actions and their perceptual consequences remain encrypted. In other words, our internal feedback systems produce unhelpful gibberish. Without predictive signaling, our behavior typically defaults to trial and error, where feedback comes *exclusively* through the external pathway, the slowest route available. As the individual blunders through this new environment, success tends to come only by chance. However, with repeated trials, such random successes contribute to a growing internal database of correlations between particular actions and their perceptual consequences in the new environment. These acquired correlations gradually improve predictive capacity. Errors diminish and the individual eventually establishes a successful control repertoire. Other theorists have also emphasized the role of learned correlations in predictive signaling (Hawkins and Blakeslee 2004).

Combined with early childhood experience, the capacity of mature adults to learn through trial and error provides a rich source of correlative data between actions and their future consequences. Depending on the situation, such advance information may represent the difference between success or failure, life or death, evolution or extinction. That such a potent capability may be a built-in feature of neural control circuits makes control architectures even more interesting as models for such cognitive processes. Further research along these lines is clearly warranted.

The Unique Human Brain

There is consensus among scholars that the brain is the fundamental organ of cognition. The evolutionary shift toward cognitive development must have begun therefore with genetic changes that affected the structure and function of ancestral brains. Scientists generally attribute the emergence of human cognition to the portions of the brain that display disproportionate expansion in the human species. As the earliest hominids diverged from their ape ancestors, each increment of new brain tissue produced an increase in information processing power. Just as modern supercomputer can hold more complex models of the world than simple desktop computers, the emerging hominid brain held mental representations of the world with greater depth, complexity, and abstraction than the simpler apelike versions. As natural selection built this new brain, it became an organ of cognition. And if we are correct in our hypothesis that cognition follows control, the organ of cognition is also an organ of control.

Physically, the vertebrate brain sits at the end of a tube running from tail to head (Bainbridge 2008). Starting at the tail and moving toward the head, the neural tube has a relatively uniform and repetitive structure, the spinal cord, which mediates sensory and motor functions through the peripheral nerves and muscles. In the earliest vertebrates, the spinal cord is the dominant structure of the central nervous system. In more advanced species, the head end of the system elaborates into a larger mass, the brain, which contains expanded cell clusters and sheets, interconnected by large bundles of filaments. Between the spinal cord and the brain, the nervous system goes through a series of transitional structures known as the brainstem.

Across species, the vertebrate brain shows progressively increasing size and complexity (Butler and Hodos 1996; Heinz, Baron, and Frahm 1988; Jerison 1973). This sequence establishes a phylogenetic continuum, which biologists sometimes consider a surrogate for the evolutionary path of the nervous system. Species whose origins appear further back in the fossil record tend to have smaller and less complex brains; later species have larger and more complex brains. The spinal cord remains fairly constant across this same continuum.

Brain architecture generally consists of a brainstem and associated cell clusters, overlaid by layered sheets and fiber bundles, known as cortex. Typically, two such cortical structures appear, the cerebellum overlaying the lower brainstem, and the neo-cortex overlaying the upper brainstem. The neo-cortex holds the greatest relevance for human cognition. Its disproportionate expansion in the human species is evident in Figure 2.4.

Advances in neuroanatomical analysis have enabled better quantification of the relative proportions of various brain structures across species (Kaas and Collins 2001). The neo-cortex in particular constitutes only about sixteen percent of the brain in primate insectivores, rising to sixty percent in new world monkeys, and finally reaching eighty percent in humans. In the lower vertebrates, the neo-cortex appears morphologically as little more than a smooth sheet that barely covers the underlying brainstem. At the other end of the spectrum, the convoluted structure of the human brain reflects a cortical sheet grown so large that it must fold and crumple just to fit inside the skull. Within the neo-cortex itself, some areas show greater relative expansion than others. The human prefrontal area, for example, has expanded at least twofold over that of the nonhuman primates (Deacon 1997; Goldberg 2001).

The volume of the human brain averages about 1300 to 1400 cubic centimeters, approximately three- to fourfold larger than that of the chimpanzee, our nearest biological relative. The human brain averages about two percent of body weight, well ahead of any other species. The human brain has more cells, more connections between cells, and greater average cell density. The human brain also shows greater metabolic activity (Leonard and Robertson 1994; Aiello and Wheeler 1995; Foley and Lee 1991; Parker 1990), consuming more than twenty percent of the body's total oxygen intake. The brains of most other vertebrates consume only two to eight percent.

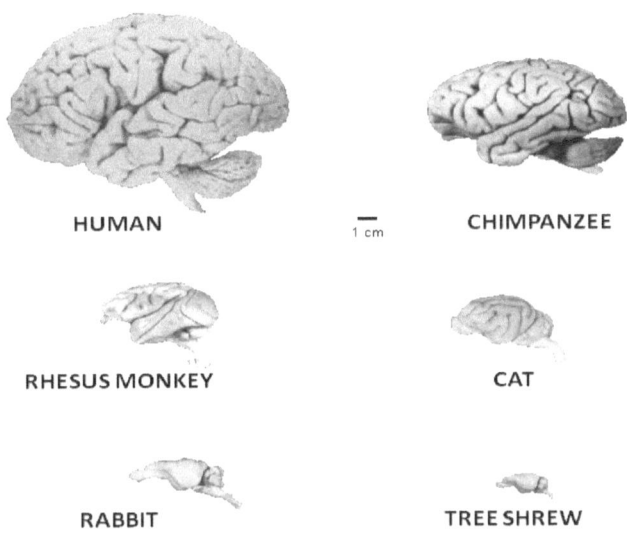

Figure 2.4. A comparative sample of external brain morphology of selected mammals. Figures are relatively to scale (1–cm key). The human brain displays a disproportionate increase in the cerebral cortex compared to the other species, including the chimpanzee, our closest genetic cousin. Adapted from http://brainmuseum.org published by the University of Wisconsin-Madison.

Much of the expansion of the human brain occurs after birth. The human neonatal brain is only about twenty-three percent of its eventual adult size, as compared to over forty percent in chimpanzees and sixty-five percent in macaques (Gould 1977). Accompanying this rapid development are wide variations between humans and chimpanzees in gene activity in brain tissues (Enard et al. 2002). In contrast, activity in blood and liver cells appear relatively similar in the two species. These data suggest that cells in the human brain utilize their genetic substrates differently than those in the chimpanzee brain.

The human neo-cortex contains about ten to twenty billion neurons, each receiving and forming about five thousand to ten thousand synapses, yielding a total synaptic population on the order of fifty to two hundred trillion (Braitenberg 1984; Koch 1998). Anatomical studies show that the vast majority of these synapses come from other cortical neurons. Neurons outside the cortex contribute only a small proportion. Braitenberg (1984) estimated that internal cortical connections dominate external by at least tenfold, possibly up to one hundredfold, as if every cortical cell were determined to connect to every other cortical cell. In effect, the cerebral cortex appears to function largely as a self-contained information exchange, where messages in one part tend to move to other parts and back.

Reciprocal connections between cortical areas dominate the anatomical picture (Felleman and Van Essen 1991), suggesting the formation of powerful looping networks. Detailed anatomical studies indicate considerable orderliness in such connections (Calvin 1996; Mountcastle 1978; Hawkins and Blakeslee (2004), suggesting that form and function obey systematic rules and relationships across the entire cortical structure.

The evolutionary expansion of the cortex in the primates, ultimately reaching its disproportionate size in the human being, lends support to the idea that the behavioral differentiation of the human species has strong links to the evolution of the cerebral cortex. As the neo-cortex surged into prominence as the human cognitive organ, it appears to have retained the same basic cellular architecture and interconnections we see in other species. In other words, the human cortex did not simply get bigger over evolutionary time; it got bigger in a precise way, maintaining its original functional blueprint, even as it expanded enormously in size. The outcome of this evolutionary shift was therefore primarily quantitative. But in its quantitative excess, this growth conveyed a qualitative change, the implications of which we are only just beginning to grasp.

Homo auguris and the Human Strategy

It is now widely believed that the extraordinary cognitive engine that is the human mind probably evolved from simpler forms, apelike and less capable, but still remarkable. To understand how and why the human species acquired this capability, we need to understand what it is and what it does. Many great thinkers have applied themselves to this task, and, perhaps predictably, many different opinions and theories have emerged. The present effort does not seek to diminish these efforts, but simply to add another idea to the mix: that the organ of cognition is also an organ of control.

By the operational definitions used in this chapter, cognition and control are not unique to the human species. Almost every vertebrate displays some capacity for both. Similarly, the putative organ of cognition and control, the cerebral cortex, is not unique. Many species have anatomically homologous structures. What makes the human species different is not the quality of our cognitive and control abilities, but the sheer quantity of processing capacity we can bring to bear. As the hominid cortex began to expand in our ancient ancestors, it retained its characteristic architecture. It continued to mediate the interplay of action and perception. But with each new increment of tissue, it acquired a capacity to represent action and perception in more detail and in more dimensions. Over time, the evolving hominids came to perceive the world differently than other species: more deeply, with a greater comprehension of its interactions, causalities, and consequences. Other species may carry rudimentary models of the world, and of themselves acting in it, but none accomplishes this feat on the human scale.

Human cognition reflects the reciprocal interplay of perception and action. It embodies both a bias for action in perceptual processing, and a bias for perception in action processing. The same control principles apply at every stage of our cognitive development, which means that every operation of the human mind, from the most elemental to the most abstract, exists in the context of control. Put simply, we perceive, act, and think with the intent to control.

I call the hominid species that first displayed these prototypical capabilities *Homo auguris,* the Seer. This hypothetical ancestor first mastered, simultaneously, both the abstractions of the world, and the abstractions of their own behavior in that world. In their challenging environment, *auguris* did not simply act on instinct and hope for the best, as their deep forest peers might. They acted as if they saw the consequences of action, in advance of action. As they gained control over more variables in their changing environment, they secured a crucial survival advantage. They were successful, not because they ran faster or climbed higher or bred more prolifically, but because they saw what other apes could not see. And increasingly, they saw it before it happened.

CHAPTER 3

Homo ipsianimus, the Self-Aware: A Circle Game

What Cognition Feels Like

In the previous chapter, I defined cognition simply as an objective instance of information flowing through the matrix of neuronal connections in the human brain, without regard to experiential consequences. The adaptive value of such systems, and hence their evolutionary emergence and preservation, depended on how well they linked information about the external environment to an individual's actions in that environment. Successful systems endured, unsuccessful ones did not. The primary hypothesis put forward in Chapter 2 was that information processing systems that utilized control architectures, that is, basic input-reference-output-feedback designs, emerged as the winners in the evolutionary cognition contest.

In the present chapter, we shall look at human cognition in a different light. Our quest here is to understand how cognitive processes give rise to the seemingly unique human experience of consciousness, that is, what thinking *feels* like. It is one of the things that humans find most interesting about themselves. It is also a very elusive concept, both scientifically and philosophically. As in Chapter 2, our goal in this chapter is to connect human consciousness to neural control systems resident in the brain. Be warned, however, this discussion will lead us to some rather strange conclusions.

Introspection

As I write, a voice in my head seems to dictate the words as I punch the keys. I recognize this voice as ME, the conscious thing that accompanies me throughout my waking day. Sometimes it speaks in English, sometimes in symbols, feelings, or visions. Its meanings usually resolve clearly and understandably, but not always. When it speaks, it seems as if I could simply open my mouth and the words would materialize, fully formed and audible. When it uses symbols and images, I feel I could just release my limbs and act in the scene before me. I cannot remember a time when the ME did not accompany me. My memory of events seems always to have two components: the event itself and the ME participating or witnessing it.

As you read these words, you may experience the same sense. It is something that each human mind seems to have, something we believe makes us unique. Understanding how it works might help us understand what makes us human. Understanding how it evolved might help us understand what propelled us out of apedom. It is a tricky problem, however.

While I know the ME in me exists, I am less certain about the YOU in you. If you and I had a conversation, I would smile and nod politely, but the ME would be saying, "You know, she could be just an unconscious machine with a clever program," and I would not know how to disprove it. Out of courtesy, I would take your word for it, that you have a YOU in you. But if the ME challenged me for scientific proof, I could muster only probability and correlation. Consciousness, the ME thing in me and (I guess) the YOU thing in you, remains utterly personal and inaccessible to outside verification. According to William Hamilton:

> Nothing has contributed more to spread obscurity over a very transparent matter, than the attempts of philosophers to define consciousness. Consciousness cannot be defined; we may be ourselves fully aware what consciousness is, but we cannot, without confusion, convey to others a definition of what we ourselves clearly apprehend [from Bowen 1863].

Stuart Sutherland (1996) goes even further:

> Consciousness is a fascinating but elusive phenomenon; it is impossible to specify what it is, what it does, or why it evolved. Nothing worth reading has been written on it.

Hamilton and Sutherland notwithstanding, dictionary writers always try to define the word consciousness. They typically express it as an awareness of one's own existence, sensations, thoughts, and surroundings. Of course, philosophers say this merely substitutes one indefinable for another: self-awareness for consciousness. The same dictionaries define self-awareness in terms of consciousness, so clearly they will not take us very far. In the following, I use the terms *consciousness* and *self-awareness* synonymously.

That the experience of self-awareness is entirely personal also poses a problem for science. Studying one's own mind provides no basis for generalization, at least not in any scientifically provable way, so every experiment involves a single sample, unconnected to others. Moreover, the scientist is also the subject, a clear conflict of interest. In order for me to study the ME, the ME must study me.

Philosophers have gotten a lot of mileage out of this situation (Chalmers 1998; Dennett 1991; Stewart and Cohen 1997), but I will not take it much further here. Instead, I shall simply assert the materialist view that everything that exists, including self-awareness, has or depends directly on a physical substrate, something tangible in the material world. If there are entities and forces outside the materialist realm, then this book (and most of modern biology) is irrelevant anyway. My construct says the ME in me owes its existence to my brain. As long as my brain functions properly, the ME exists. It produces words, images, and symbols that I recognize and understand as a conscious being. If my brain changes significantly, for example, from injury, disease, drugs, toxins, sleep, or stress, then my state of consciousness should also change. To the degree that you have a brain similar to mine, the materialist construct says that you too will have a consciousness like mine, assuming that causality works the same in your neighborhood as in mine. In other words, the YOU in you has properties similar to the ME in me. Both derive from our similar brain structures and neural functions.

Measuring self-awareness remains a problem under the materialist construct, but some surrogates have proven useful. Take, for example, the Turing test. In 1950, mathematician Alan Turing formulated a conceptual protocol to address the subjective problem of intelligence in machines (Turing 1950; Saygin, Cicekli, and Akman 2000). His test involves a Judge and a Candidate who exchange messages through a neutral interface, like a computer terminal. They have no other direct contact with each other, and neither has any specific prior information about the other. In an open-ended process, the Judge and the Candidate exchange messages until the Judge determines whether the candidate has the particular subjective property under consideration. In Turing's original problem, the Judge would unknowingly exchange messages with a computer running an advanced program. The Judge's verdict, that is, whether the Candidate had intelligence or not, would determine whether the computer designers had succeeded in building an intelligent machine.

In principle, we can apply the Turing test to any problem with an irreducible subjective component. Research on self-awareness often incorporates this approach, although usually not called out as such. The investigator (Judge) asks questions, makes statements, or gives commands, and the subject (Candidate) answers or behaves in some way. The investigator then assesses whether the subject meets some criterion for self-awareness. The investigator might also apply an experimental procedure or make measurements during the exchange, for example, a brain scan that measures patterns of neural activity. This might reveal certain physiological correlates with the inferred self-awareness. The experimenter has no direct access to the phenomenon, just as Turing had no direct access to a machine's intelligence, but the protocol gives everyone an agreed starting point.

Gordon Gallup and colleagues (Gallup 1970; Keenan 2003) took this approach a step further with their famous mirror test. Test subjects, which included human adults, children, apes, monkeys, or other species, were placed in a laboratory room with a mirror. For some subjects, like chimpanzees and monkeys, this was their first exposure to mirrors. They displayed diverse reactions, often showing great fascination with their own reflections. After a few days, the experimenter surreptitiously marked the subject's forehead with a colored dot, taking care not to reveal to the subject what was going on. The subjects returned to the mirror room, and the experimenter again observed their behavior. Some subjects, specifically human adults, older children, and a portion of chimpanzees and other apes, immediately touched *their own foreheads* as they looked at their now-altered images in the mirror. The others, children under the age of two, monkeys, and other species, did not touch the spot. Gallup hypothesized that touching one's own body in response to an external reflection indicated self-awareness. Although indirect and inferential, the mirror test is a clever way to identify those individuals and species that behave *as if* they had an awareness of self. The fact that only humans and a portion of the great apes respond reliably in this test suggests the relative enhancement of self-awareness in these species over that of others.

Gallup's results suggest two things about self-awareness. First, it is not necessarily unique to humans, but second, it may be relatively rare across the animal kingdom, at least in this measurable form. Thus, the potent sense of self-awareness in humans may be a quantitative distinction rather than a qualitative one. If it emerges from brain structures, as the materialist view requires, then our search for its underlying mechanism should perhaps focus on areas of quantitative enrichment in the human brain. Fortunately, we have decades of research in comparative neuroanatomy and neurophysiology to draw upon, once we decide what to look for. But deciding what to look for is not so easy.

The Circular Logic of Self-Awareness

If, in our materialist view, self-awareness emerges from the interaction of neurons arranged in particular configurations, then our first challenge is to determine which arrangements might have this property. If we can then show that such arrangements exist, and that they are more prevalent in human brains than in other species, we will have taken a step toward specifying the evolutionary path from apedom to *Homo ipsianimus*, the first self-aware hominid.

Let's start with the basic observation that self-awareness has a circular quality. Like holding a mirror up to another mirror, the human mind can seemingly generate an infinite series of self-reflections. If I am self-aware, I am aware of my own self-awareness. I am, in turn, aware of my own awareness of my own self-awareness, and so on. The neural system that encodes my mind seems able to generate and recognize representations of itself. Each turn around my self-awareness loop seems to create a new version of my own representation until eventually I run out of mental resources, or I get bored and stop. Douglas Hofstadter (2007) recently explored the circularity of self-awareness in his remarkable analysis of *strange loops*.

If circular logic is a characteristic of self-awareness, then clearly we should look at signaling arrangements that incorporate circularity as candidates for mediating the property of self-awareness. Consider, for example, the schematics in Figure 3.1. Each panel shows short signaling chain consisting of two elements, *A* and *B*. These elements can be anything: people, mice, neurons, transistors, and so on. The signals can also be anything: voice conversations, distress calls, chemical releases, electric currents, and so on. The signal enters at the left, engages elements *A* and *B* in some way, then exits at the right.

In the first chain, the incoming signal engages element *A* which in turn engages element *B*, which in turn passes the signal downstream. Element *A* clearly has an effect on *B*, but *B* has no effect on *A*. If we could precisely measure the activities of elements *A* and *B*, we should see evidence of *A* encoded in the activity of *B*, but not the other way around.

Now compare the second chain. Here, element *A* engages element *B*, as before, and *B* engages downstream elements, also as before. However, in this case, *B* also sends a return connection back to *A*. We call this a recurrent loop. Element *A* receives this recurrent signal as new input, transforms it in some way, then sends a new signal to *B*, which in turn sends new signals downstream. But *B* again sends a return signal back to *A*, and the whole process repeats. The loop causes signals to move repeatedly back and forth between *A* and *B*. If we measure the activities of elements *A* and *B*, we should see not only the effects of *A* appearing in the activity of *B*, but also the effects of *B* appearing in the activity of *A*.

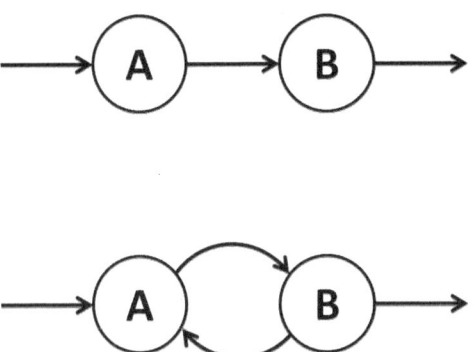

Figure 3.1. Two simple signaling chains with very different properties. *A* and *B* can be any entity capable of sending and receiving signals. In the first chain, the signal enters from the left, activates *A*, which in turn activates *B*, which sends its signal onward. If we define awareness as a change in state or behavior due to the detection of a signal, then *B* is aware of *A*, but not the other way around. The second chain incorporates all the same features as the first, but it adds a feedback signal from *B* to *A*. Now both *A* and *B* are aware of each other. More importantly, because each now also detects its own prior signals sent to the other, each is now also aware of itself. This simple model suggests that self-awareness is a topological property of all looped feedback systems.

Now let's carefully overlay the concept of awareness. Here we define awareness in purely operational terms: specifically, that awareness reflects a change in activity in one element caused by the detection of signals coming from another element. With this overlay, how do our previous descriptions change? In the first chain, we would say that *B* has awareness of *A*, but not *vice versa*. Element *A* does not detect any signal from *B*, so it is not aware of *B*, by our operational definition. In the second panel, elements *A* and *B* show mutual awareness. Each detects a signal from the other, so each has awareness of the other, again according to our operational definition.

Now here is the important part. *The loop structure creates the logical property of self-awareness.* Element *A* only becomes aware of *B* after *B* sends its return signal. But *B* cannot send its return signal until *A* transmits its original signal. Thus, when *A* detects the return signal from *B*, it also detects *its own prior signal* to *B*. In effect, *A* detects a signal from *A*. By our definition,

A becomes aware of *A*, that is, self-aware. Similarly, *B* initially has awareness only of *A*, but when *A* resends its signal after receiving feedback from *B*, *B* also becomes aware of *its own prior signal* to *A*, and it too becomes self-aware. Thus, both elements in this loop arrangement behave as if they had awareness of each other *and* awareness of themselves.

Let's apply this logic to a specific situation. Suppose we define *A* and *B* as people talking on the telephone. In the first case, *A* speaks to *B*, but *B* does not speak to *A*. We know that *B* has awareness of *A*, but *A* does not necessarily have awareness of *B*. Person *A* cannot determine whether person *B* exists or not, since *B* has not spoken. In the second case, both *A* and *B* are aware of each other: *A* said something and *B* said something in reply. Moreover, if *B*'s reply has some correlation with *A*'s initial message, *A* knows that *B* heard the initial message, as opposed to just some random noise. Person *B* also knows that *A* heard the feedback reply, provided that *A*'s reply has some correlation with *B*'s reply. Thus, both *A* and *B* have awareness of each other, but they are also aware of their own prior messages in the conversation, and therefore their own presence in the system. At any given moment in the ongoing exchange, both parties receive awareness information about each other, and about themselves.

Now, defining *A* and *B* as people having a phone conversation makes the inference of human self-awareness more intuitive. But let's take a bolder step. Suppose we define *A* and *B* as individual neurons, and the signals between them as excitation or inhibition through the release of neurotransmitters. Does the situation change? Logically, the same patterns apply. In the first case, neuron *B*'s activity reflects the activity of *A*, but not the reverse, so we can say neuron *B* is aware of *A*, but not the other way around. In the second case, each neuron detects the activity of the other, so each is aware of the other, but each also detects information about itself at a previous moment, which, by our operational definition, constitutes self-awareness. Thus, we cannot escape the logical conclusion that self-awareness exists at the neuronal level. Neurons that participate in recurrent signaling loops must encode, in part, their own presence in the system.

Some readers might balk at this remarkable conclusion. How can a single cell possess what we humans experience as self-awareness? The answer is simply that experience is not the relevant point. We can also say that humans experience life, but we cannot then use this to deny that individual cells have life. Life is a property both of cells and of human beings. How each experiences it is a different question. Similarly, self-awareness is a property both of neural loops and of human beings. How each experiences it is again a separate question.

Now let's extend this analysis with a thought experiment. Starting with the basic structures in Figure 3.1, let's construct all possible combinations of excitatory and inhibitory neurons. Then let's multiply each combination several millionfold and connect them together in various ways until we build a replica of a human nervous system. According to the materialist view, at some point in this construction, we must produce a system that experiences human self-awareness. The question is when. How many elements, arranged in what particular structures, produce true self-awareness? A million? Ten billion? As you ponder your answer, consider this. If you pick any number greater than two, you must explain why that number yields self-awareness, while systems with one less neuron do not.

My answer is two, and the requisite structure is an interactive loop. Self-awareness is simply a topological property of all interactive loops regardless of their substrates, biological or nonbiological. Recurrent architectures, no matter what size, complexity, or chemical composition, define self-awareness. To find it, you simply look for signaling loops.

In the simplest system, the single recurrent loop in Figure 3.1, the recurrent interaction encodes the totality of its self-awareness. It is meager, inaccessible, and utterly unfamiliar to our human experience, but it exists nonetheless. Each such loop, when combined in the billions in a complex nervous system, has the same self-awareness as when it stands alone. In larger systems, however, each bit of self-awareness connects to many other bits, so that the whole system has an aggregate self-awareness comprising many billions of individual bits. It is in such large and complex systems that self-awareness becomes more recognizable by human experiential standards. By analogy with physics, large-scale self-awareness emerges from neural self-awareness in the same way that Newtonian mechanics emerges from quantum mechanics. Every recurrent neural interaction contributes a momentary quantum of self-awareness to the system. Newtonian self-awareness results from the superposition of many such quanta.

By this reasoning, the experience of human self-awareness derives from the actions of billions of interactive loops, each achieving a brief logical state of self-awareness during their local recurrent interactions. Every neuron that participates in a recurrent signaling loop encodes both self and non-self, simultaneously and inseparably. Thus, if our brains experience anything at all, they must also experience self-awareness.

Defining self-awareness in this way changes the problem. It is not an exclusive human property, but a ubiquitous phenomenon with many manifestations, both biological and nonbiological. We may not always recognize it as such. When we see two people talking to each other, we find it easy to interpret their expressions and behaviors as reflecting individual self-awareness, as we understand it. When we see a chimpanzee touch its own forehead in response to a surprising reflection in a mirror, we can easily attribute some sense of self. In contrast, when we measure two interacting neurons, we find it harder to draw the same conclusion. Yet logically, they are the same, except for the obvious differences in scale and mode of expression. If human self-awareness lies in brain structures, and brain structures build on interacting neurons, then the question always comes down to where in the aggregate does experience change from quantum nothingness into a recognizable ME.

If self-awareness emerges from recurrent interactions, then it follows that humans have a greater sense of it simply because our brains have a much richer set of recurrent interactions than other species. If other species have nervous systems built on recurrent interactions, they too must have self-awareness, just not to the same expressible degree as humans. Logically, even nonliving systems, like computers and robots that incorporate recurrent feedback circuits, must also have the property. We cannot access it, and they cannot express it, but it must be present in all such circuits, regardless of their nonbiological status. Just as we cannot prove consciousness in other humans, we cannot disprove it in machines. After all, in the materialist view, we are all machines.

The Circular Logic of Control

If self-awareness is a topological property of interactive loops, then our search for *Homo ipsianimus* becomes a search for neural systems that have the requisite loop architecture. We can be confident, within the assumptions and limits of this analysis, that such systems will contribute to the self-awareness we all experience in our waking lives. As you might have guessed, we need look no further than control theory to find such structures. Indeed, recurrent signaling is one of the defining features of control.

Of the various building blocks developed by control theorists, *feedback* represents the classic model. Feedback control involves connecting the system's output back to its input, as suggested in Figure 3.1. Systems can use positive or negative feedback or both. Positive feedback uses the initial output to amplify the input signal, thus increasing the output still further. This type of feedback typically achieves the desired goals of rapid growth, self-organization, and signal amplification. Negative feedback uses the initial output to suppress the input, thus tending to reduce subsequent output.

This type of feedback typically achieves the goals of stability or controlled oscillation. Clearly, all such feedback mechanisms incorporate the feature of circularity. Loop structures inextricably mix output and input, so that each reflects the other, and each encodes its own prior action. All such feedback control circuits must therefore possess the logical property of self-awareness, as we have argued above.

Taking the next logical step, we can build larger, more complex control systems such as the one used in the previous chapter (see Figure 2.2). In principle, each of nested loops within this architecture must contribute to the system's aggregate self-awareness. Multiplied several billionfold, such complex control architectures could serve as the physical substrates of self-awareness. Thus, the same neural structures that mediate adaptive information processing can be simultaneously a rich source of self-awareness. If the human nervous system incorporates such neural control circuits in sufficient quantity, then we will have found a prime candidate for both human cognition and self-awareness.

Looking for Loops

Neuroscientists have long believed that recurrent cortical networks have special importance in human mental functions. In the late 1940s, theorists led by Donald Hebb (1949) first developed the concept of the *cell assembly* in which groups of neurons self-organized into recurrent loops, such that activating any part of the loop would activate the rest. The self-excitatory feedback structure of the assembly would allow it to sustain its activity for some time, even after the primary stimulus had stopped. In Hebb's view, *thinking* represented simply the sequential activation of cell assemblies.

With the advent of modern experimental techniques and theories, Hebb's original idea that neural loops played a role in higher brain function has received renewed attention. For example, Gerald Edelman (1992, 2006) has mustered an impressive array of logic, argument, and evidence to support the hypothesis that looped or *re-entrant* cortical signaling establishes functional neuronal groupings that mediate mental states and define behavioral outcomes. Francis Crick (1994) has likewise noted the importance of looping pathways in understanding awareness, focusing primarily on the visual system. Crick suggests that recurrent networks involving both cortical and subcortical centers may play an important role in mediating consciousness.

In the clinical domain, neurologists believe that self-exciting cortical loops are the basis for certain forms of epilepsy, the clinical condition that produces debilitating seizures (Penfield and Jasper 1954; Delgado-Escueta et al. 1999). Epileptic pathologies arise from the spontaneous activation of a specific point in the loop, which progressively entrains other connected areas, until the entire network of loops ignites in sustained volleys of self-excitation. Depending on how many loops ignite, the behavioral consequences of epileptic seizures range from mild to catastrophic.

Not all neural loops degenerate into epileptic pathologies, of course. The reason is that they are actually well-controlled systems that incorporate feedback inhibition in addition to feedback excitation. Neurophysiological studies reveal large numbers of both inhibitory and excitatory neurons in all brain areas, suggesting that normal recurrent neural systems operate under precise control. Theorists have suggested that control mechanisms must be integral to cell assemblies in order to preserve their utility as adaptive behavioral mechanisms (Braitenberg 1984). When disease or pathology disturbs this control balance, the system may degenerate into the chaotic functioning characteristic of epilepsy.

The prevalence of recurrent structures, containing both positive and negative feedback, suggests that the human brain, particularly the cerebral cortex, functions both as a powerful behavioral control system and as a strong focus of experiential self-awareness. Thus, the ME in me and the YOU in you are functions of ubiquitous looped structures located throughout our cortices and the rest of our brains. While our experience of self-awareness may have no real relevance to the underlying function of cortical loops, it gives us the sense that something special is happening in our heads, something that other species do not seem to have to the same degree.

The Equivalence Principle

We can summarize the main idea of this chapter in an equivalence principle. Recurrent signaling, control, and self-awareness are not independent entities or processes, but rather coequal descriptors of the same underlying phenomenon. Recurrence describes the physical layer, control describes functional relationships, and self-awareness describes the epiphenomenal experience. While each has its own tradition of scientific and philosophical study, as if it somehow had an independent existence, all are in fact embedded properties of one another.

The equivalence principle says that we cannot separate the feeling of self-awareness from the functional outcomes of control processing, or from the signaling properties of circular networks. It cannot exist without them, and it cannot be suppressed if they are present and operating normally. Self-awareness is simply *what control feels like.* Control, in turn, is simply the outcome of particular forms of recurrent signaling. The seeming augmentation of self-awareness in humans must therefore derive from increased populations of recurrent neural structures in our brains. As loop circuits began to proliferate in the brains of the ancient ancestral apes, both the cognitive ape and the self-aware ape began to emerge. Both traits reflect the selective advantages of a brain growing larger in a very specific way.

On Emotions

There is another aspect of self-awareness that requires some discussion in this chapter: specifically, that self-awareness seems to come in different flavors and colors (speaking figuratively, of course). These varieties of self-awareness are the emotions, and much has been made of them as an essential part of the human psyche (Damasio 1999; LeDoux 1996; Greenspan and Shanker 2004). To some theorists they are crucial, indeed the very heart of humanness. From the experiential perspective, this may be true. The emotions surely add a rich tapestry of colors and textures to our conscious lives. However, from the operational and evolutionary perspectives, the importance of emotions as experiential events is less clear. As we have argued above, the experience of self-awareness is a property of recurrent signaling loops in the brain. It follows that the different varieties of self-awareness that we call emotions are simply reflections of different populations of recurrent control loops becoming active at any given moment in any given set of circumstances. As the active population of loops varies, the quality of self-awareness also varies. To each distinct and replicable set of active loops, we may attach a label, such as anger, fear, happiness, sadness, disgust, and so on.

The American Psychological Association (APA) defines emotions as complex reaction patterns, involving experiential, behavioral, and physiological elements, by which the individual attempts to deal with a personally significant matter or event. Note that the three basic elements—physiological, behavioral, and experiential—are the same elements postulated above in the equivalence principle. The APA definition adds that emotions arise without conscious effort, and have either positive or negative affect. Thus, according to the standard definition, emotions arise as interactions within functional neural arrangements that convey specific affective experiences, depending on which elements or arrangements happen to be active at the moment. This clearly looks very similar to our general definition of self-awareness above, with the added feature of differential affect.

Marvin Minsky (2006), one of the founders of artificial intelligence and a leading light in cognitive science, recently offered a more basic definition of the emotions as simply "different ways to think." Minsky suggests that as the brain mobilizes cognitive resources in response to particular situations and problems, their aggregate patterns carry different affective properties, to which we may assign different emotional labels. Operationally, however, they are simply different sets of interacting brain resources engaged in different kinds of thinking. Again, the fact that they give rise to different feelings is interesting, but not necessarily the most evolutionarily relevant property of the system.

Minsky's notion of different ways to think obviously fits with the basic thesis of the present work. In Chapter 2, I suggested that thinking is the operation of large arrays of neural control modules. In the present chapter, I suggest that these same thinking arrays, to the degree they are looped structures, must also give rise to the sense of self-awareness. If different modes of thinking imply the activation of different control arrays, and if the activation of different control arrays gives rise to different affective qualities of experience, then the present model complies precisely with both the APA definition of emotions outlined above and with Minsky's more elegant assertion.

The crucial question about emotions in the present context is not what they are precisely, but what they indicate about the adaptive value of cognition and self-awareness. We can ask this question in two ways. First, is the *experience* of emotion a decisive factor in determining the adaptive value of the trait? The answer, I suggest, is an unqualified no. Natural selection does not favor us merely because we are capable of feeling anger, or happiness, or even love. In the crucible of evolution, emotional experiences are irrelevant unless they have a material effect on the outcome of natural selection. Feeling good or bad about things does not necessarily translate into selective fitness.

The second way to ask the question is functional. Does the presence of differing emotional states materially influence relevant behavioral outcomes under the prevailing selective conditions? In other words, does the system that produces emotions also produce adaptive actions that improve an individual's survival chances and/or reproductive opportunities in a given environment? Here, I believe, the answer is a qualified yes. If there is strong selective pressure in favor of certain functional systems, and those systems also generate certain emotional experiences, then we can say that the emotions are at least indicators of important and evolutionarily relevant events. Of course, their adaptive value lies entirely in their functional outcomes, not in their experiential correlates. Again, nature does not reward us for feeling but for doing.

What then are the adaptive correlates of the emotions? From a purely biological perspective, almost everyone would agree that the adaptive value of the emotions lie primarily in their motivational properties. Emotions carry an implicit call to action, either in a positive context (e.g., joy) or in a negative one (e.g., anger). If we believe that the various emotions arise from the differential activation of various sets of control loops, then it follows that their motivating qualities must reflect the differential functions and activities of such control loops. Referring back to Figure 1.1, which shows the basic architecture of control, the primary element associated with motivation is *dissonance*. As we have outlined previously, dissonance is a pattern of neural activity that arises from the mismatch of perception and expectation. Its operational role is to trigger actions that alter perception, until the mismatch disappears. Dissonance is therefore the primary motivator of action in organisms that utilize neural control systems. The obvious question is whether the various patterns of dissonance that arise in complex networks of control loops contribute to the experiential phenomena we know at the emotions. This is a very large question. We have no definitive answer yet.

In common English usage, the term *dissonance* has a generally negative tone. It implies something is wrong or out of balance, which in our control model is precisely true. Dissonance represents a state of relative unease from which some form of behavioral striving must originate. From this starting point, we could hypothesize that dissonance patterns correlate primarily with negative emotions. We might therefore assign to such patterns various emotion words, such as like anger, yearning, fear, sorrow, and so on, depending on their particular origins and associated actions. Regardless of their labels, however, dissonance patterns still retain their functional identities as fundamental elements in control systems.

The logical converse of the above hypothesis is that the reduction of dissonance should therefore correlate with the emergence of various positive emotions. In a cognitive control system, successful action brings about a new set of perceptions that null the mismatch between previous perceptions and the current expectations. This reduces dissonance, which, in the present context, means it removes a negative emotion. The experiential result may be an affective reversal from negative to positive: anger gives way to joy, yearning to contentment, fear to relief, sorrow to happiness, and so on. Positive emotions therefore represent the emergence of various null states in complex cognitive control systems.

As the human control network churns through its daily programs, there is a constant ebb and flow of dissonance, reflecting a complex system engaged in its different modes of thinking. Each of these modes carries an element of self-awareness to which we may attach an emotional label, according to Minsky's definition. The qualities and intensities of such emotions depend on the precise patterns of dissonance, on how they arise through cognitive mismatches, and on how they dissipate when actions, both overt and covert, alter the perceptual mix.

While this simple control view of human emotions may not satisfy many researchers in the field, it does provide a plausible operational foundation for how emotions might arise. It also makes a rational argument for why humans seem to have a greater emotional range than other species (i.e., more control loops). Finally, it suggests a basis for how the emotions evolved as part of the human cognitive structure. Obviously, much further study is needed.

Homo ipsianimus and the Human Strategy

Our imaginary ancestor, *Homo ipsianimus*, the Self-Aware, was the first to experience that inkling of self-awareness now so familiar to us in our waking lives. We came to this state not because self-awareness necessarily has any direct adaptive value, but because it reflects the underlying engine of human evolution: a large cerebral cortex organized in a massively recurrent control architecture. The cortex takes in information about the world, compares it to its internal templates, and generates complex behavior to make perception conform to expectation. As this cortical factory churns through its program, each pass through a recurrent loop contributes a flicker of self-awareness to the aggregate. With so many flickers, the flame of self-awareness burns brightly in our species. But it is only that, a bright flame that sits on top of the powerhouse. The real work is done below.

Every recurrent structure, both living and nonliving, has that same flicker of self-awareness. In machines, it may exist only momentarily when feedback control mechanisms kick in. We cannot access it and they cannot express it. The fact is, we cannot access it even in other human beings, yet we still believe it exists, and that it says something special about us as a species. Of course, there is no one to tell us otherwise.

Theorists disagree on the timeline of human self-awareness, so we cannot say whether *Homo ipsianimus* is ancient or relatively modern. Most paleontologists believe that by the time of *Homo erectus*, one to two million years ago, the hominid line had achieved a level of self-awareness we would recognize as human. Whether it was present earlier, we cannot say. If we accept the primate research, we must grant the chimpanzees at least some sense of self-awareness, which means that it may have been present fairly early in the hominid divergence. If we accept the neo-cortex as the essential neurological differentiator, then the path toward human self-awareness should parallel cortical evolution. A large cerebral cortex would have given *Homo ipsianimus* a stronger predisposition to control events, and a growing confidence to succeed in hostile new territories and harsher habitats. In time, it would also have given them the ability to reflect on themselves, and to sing their own praises.

Regardless of when it started, self-awareness exists in modern humans as a familiar daily companion. As we accomplish our mundane tasks, many of them truly remarkable when measured against our humble evolutionary origins, we too have the ability to reflect on ourselves and to sing our own praises. In that, we are truly the descendents of *Homo ipsianimus*.

CHAPTER 4

Homo conlocutus, the Converser: Sharing Thoughts

Are You Talkin' to Me?

In the famous mirror scene in the film *Taxi Driver*, actor Robert De Niro shows us a character, Travis Bickle, reaching terminal velocity in his descent into alienation. Unable to connect with his world, he resorts to verbal abuse of his mirror image. He says, "Are you talkin' to me? Well, I'm the only one here." We hear only his responses to the mirror, almost comical in their pointless abuse. We do not hear the other side of the exchange. We can only guess whether the character does. In the end, of course, things go badly for him.

It is a disturbing scene, in part because of character's lurking violence, but also because of the way he uses language. Not that Travis Bickle's grammar and pronunciation are incorrect. They conform to standard usage, at least for the New York streets. He speaks clearly and we understand what he means. His behavior disturbs us because he takes language out of its true human context. Language is about *conversation*, the exchanges that take place between people. Words and syntax connect us. Their interplay binds us together, transforms us, and creates new mental states and behaviors. When they are done, they release us to go our independent ways, but always changed to some degree on some human dimension. Travis Bickle leaves out the essential ingredient of language: another person. His *conversation* is not a path to connectedness but a short circuit to oblivion. The more he presses, the more he distorts his own perceptions, eventually spiraling out of control. In this chapter, we explore how language works in the context of human control, or in Bickle's case, how it should have worked.

The Adaptive Value of Language

Nearly all species engage in some form of communication. From modern primates to ancient organisms, signaling utilizes virtually every practical modality: acoustic vibration, electromagnetic energy, tactile manipulation, biochemical exchanges, and so on. Some biologists consider communication one of the defining properties of life, that without such exchanges, complex organisms could never have emerged from the primordial ooze.

At its most basic, communication connects a sender and a receiver in a relationship that alters the receiver's actions or internal states. In any given situation, individuals may trade roles as sender and receiver, sometimes repeatedly. Their mutual transformation may be subtle or profound. Its effects may last a few seconds or a lifetime. It may occur within species or between species, in some cases producing a mutually adaptive result for both sender and receiver, in others leading to a distinctly maladaptive outcome for one or the other.

The science of communication was pioneered by Claude Shannon as a prelude to the cybernetic revolution (Shannon 1948; Shannon and Weaver 1949). His work characterized communication in terms of information flow, relating it to the deeper mathematical logic of entropy and probability. Communications theory analyzes how the flow of information transforms cybernetic states.

Of all the forms of communication, human language is seemingly unique in its complexity, sophistication, and power. Some argue that language is the one true defining human characteristic. If that is so, then we must find a way to connect this remarkable phenomenon to control, the fundamental theme of this book.

In this chapter, I will build a case for human language as the primary means by which individual human control systems connect with one another to create larger multi-unit control systems that serve the shared expectations of human organizations. In other words, language evolved to synchronize controlling minds into a larger controlling cooperative. The adaptive value of such a process seems obvious. Collective action applied cooperatively to a shared purpose has a higher likelihood of success than individual action alone or in the uncoordinated context of a mob. The power of numbers augments the power of individuals. Language provides a means for creating, shaping, and directing this cooperative power.

Consider the following scenario. A group of hominids lives in a wooded area drained by a swift flowing river. The group occupies the territory on one side of the river. It is rich enough to sustain them, but access to the other side would open a new range with even more resources. That would surely mean improved health, survival, and reproductive success. The problem is they cannot safely cross the stream without some kind of bridge. Let's suppose that there is sufficient fallen timber in the vicinity, and that it is large enough to create a serviceable bridge, if only it were in position to span the stream. How can the group make this happen? There are three options.

First, each individual could proceed independently and attempt to build a personal bridge. Let's suppose this approach invariably fails because no individual acting alone can move logs large enough to span the stream.

In the second option, the members of the group could act as a mob without cooperation or plan. The massed group converges on a log and attempts to move it. The added force of numbers offers more potential power to move the log. However, because the group members each have their own personal control agendas (i.e., perceptions and expectations), they tend to work at cross-purposes, interfering with one another until the combined effort degenerates in futility. Although the mob is physically more powerful than the individual, its lack of focus and coordination represents little more than a hodgepodge of individual efforts.

The third approach, true cooperation, is more promising. If the group can somehow establish a shared cognitive representation of the situation in the mind of every group member, then the outcomes of their actions may be more effective. If the participants all see the same problem in the same way, that is, they have the same set of expectations and perceptions, they should all have the same propensity for action, assuming some similarity in their brain structures and their prior experiences. By establishing a shared agenda, the group should be able to work together more effectively. Once every individual is inclined to act in roughly the same way, the group as a whole should experience less interference and conflict, and should therefore be better able to combine their physical power to overcome the inertia of the logs. The clearer the group's shared representation of the situation, the more cooperative its actions will appear.

How can the group achieve this shared agenda? The answer, I believe, is syntactic language. By conversing with one another, individuals can replicate their individual cognitive states in other brains. The resulting cognitive synchronization among individual control systems generates a multi-unit control organization, which is capable of directing greater force and power to solve shared problems. Over evolutionary time, this trait should translate into a selective advantage. The balance of this chapter explores how this synchronization might happen.

Note that I am not suggesting any *group consciousness* at work here. The notion of a multi-unit control system does not imply individuals transformed into mindless robots performing fixed action patterns. In cooperative organizations, individuals still act independently, based on their own personal perceptions measured against their own personal expectations. The resolution of personal dissonance is still the primary measure of success. The multi-unit control system is simply an adaptive social structure in which each individual, functioning independently, comes to see cooperative action as having the best projected personal outcome.

Once they complete the action, and individual dissonance has dissipated, the control cooperative may simply disband. Participants go their separate ways to fulfill other individual expectations, until, at some future time, they may reconstitute a new cognitive organization to address a new problem. Humans move effortlessly between these two states, individual and organizational, the former usually characterized by quiet independence, the latter by a marked increase in chatter and gesturing. Language is the remarkable agent that entrains otherwise independent cognitive control systems into multi-unit cooperatives with better prospects for success.

The name I have coined for the first hominids to use language to solve daily challenges is *Homo conlocutus*, the Converser. Their unique form of communication developed as a natural extension of their enhanced cognitive development. This in turn arose from the augmentation of control architectures in the neuroanatomy of their brains. With syntactic language, *Homo conlocutus* acquired the capacity to replicate cognitive states, however rudimentary at first, from one individual to another. As a result, their prospects for survival and reproduction improved, which in turn perpetuated the genetic foundations both of their emerging cognitive minds and of their ability to synchronize those minds through language. The topic of their first conversations was probably the same as most of our conversations today: how to gain and maintain control over the events that affected their lives.

Linguistics and Universal Grammar

Linguistic science defines a language as an arbitrary system of phonetic and semantic elements that combine according to a set of rules to transmit thoughts, feelings, and information from a sender to a receiver. Typically the phonetic elements, in the form of sounds or gestures, combine into a lexicon or catalog of smaller units, words, which assemble into larger units, sentences, according to the language's rules or grammar. Sentences typically acquire semantic properties, meanings, both from their component words and from the sequential arrangement of those words. Both verbal and nonverbal languages (e.g., American Sign Language) have precise structures, but one uses sounds while the other utilizes physical gestures and facial expressions. Most languages also have written forms, equally constrained by grammatical rules, which utilize static visual symbols arranged in some pattern. Nearly every person on earth has language capacity. The Ethnologue database estimates about 6,800 living languages that vary from a few dozen speakers to hundreds of millions (Grimes 2000).

Linguistics encompasses the study of language, including its structures, expressive symbols, and meanings, as well as the place of language in human behavior and culture. Linguistics originated in the eighteenth century as a branch of philosophy. Its scientific foundations built on comparative studies of language families such as the Indo-European. In the early part of the twentieth century, the concept of structural or descriptive linguistics redefined the study of language as a systematic structure linking thought and symbol (de Saussure 1916; Robins 1979). There followed in the 1950s the analytical approach known as transformational-generative linguistics, which continues as a powerful force in modern linguistic science (Chomsky 1957).

Linguistics has traditionally focused more on the forms and structures of language, and less on its biological implications and evolution. Language leaves no fossil traces, so evolutionary models require considerable speculation. More recently, however, as neuroscience and cognitive science have developed better models of the brain and mind, and as paleontology has produced a deeper understanding of hominid evolutionary forces, the study of language evolution and its functional role in human behavior has attracted greater interest. Several prominent linguistic theorists have engaged the biological and evolutionary foundations of language (Lieberman 1984; Pinker 1994; Bickerton 1995; Calvin and Bickerton 2000; Jackendoff 2002; Greenspan and Shanker 2004; Kenneally 2007).

Noam Chomsky, the central figure of modern linguistics, opened the door to biological interpretations of natural language with his theory of universal grammar (Chomsky 1957). Chomsky's thesis pivots on a fundamental observation: Language development does not follow a completely independent course across diverse human groups; it invariably conforms to certain basic patterns and constructions. This is a profound observation. If languages were built entirely as free-form cultural inventions, we should expect human groups, isolated by time and distance, to create widely disparate languages, with unique syntaxes, phrase forms, and semantic conventions, each virtually untranslatable to the others. In fact, this does not happen. All natural languages show certain structural similarities, as if in all the collective human imaginings on how to put a language together, only a relative few make sense. Chomsky reasoned that humans must have a predisposition toward certain syntactical patterns out of the infinite variety possible. He referred to this underlying linguistic skeleton as the *universal grammar*. On this foundation, speakers build the basic elements of all natural languages, including the enormous variety of words, phrases, and sentences that characterize any modern language.

Chomsky's theory revolutionized thinking both in linguistics and in the cognitive sciences. Prior to this, theorists tended to favor the *tabula rasa* or *blank slate* view of human development. This philosophy holds that humans are born as largely unprogrammed machines, and that cultural experience supplies all the necessary information to program the cognitive mind. Carried to its extreme, the *tabula rasa* theory implied that cultures could manufacture any arbitrary language, which their children would acquire and speak as fluently as any other. Chomsky's theory shot a gaping hole in this view. Language stays within particular constraints even across diverse cultures. Infants and toddlers instinctively ignore linguistic constructions that fall outside these constraints. Chomsky's view implied that language had powerful innate determinants. Its proper foundation lay closer to biology than to pure cultural invention or raw behaviorism.

While universal grammar remains a compelling, if sometimes controversial, hypothesis in modern linguistics, its evolutionary origins and neural foundations remain uncertain. Modern neuroscience drew the conclusion that the developing human brain must have preconnected neural assemblies or modules that underlie the acquisition and expression of language. Some scholars see the language module as a recent evolutionary development, present only in modern humans using advanced syntactic language. Others see its antecedents as quite ancient, something that set the earliest hominids apart from the other apes. If it is an early feature, it most likely did not support advanced grammar, as the earliest hominids probably did not communicate in complete grammatical languages. On the other hand, while they may not have had true human language at the start, the early hominids must have reached a point in their evolution when they no longer communicated in the traditional ape fashion either. Perhaps the mechanisms of universal grammar represent the organizing templates for these original attempts at non-apelike communication. As the hominid divergence progressed, these templates evolved hand in hand with the human cognitive engine, until they provided linguistic organizing functions at a much more advanced level. As these language mechanisms evolved from ancient hominid to modern human, they may have left their traces as *linguistic fossils* (Jackendoff 2002) embedded in the strata of modern language and culture. The following section outlines one possible approach to uncovering such fossils.

Stephen G. Dennis

Protolanguage

Derek Bickerton has emerged as one of the major linguistic theorists pursuing the origins and functions of language in the human species (Bickerton 1990, 1995, 2008; Calvin and Bickerton 2000). Among many contributions, Bickerton introduced the concept of *protolanguage*, a degenerative form of natural language with fewer syntactical rules and reduced utility. Protolanguage results when the brain initiates the development of language, based on its innate neural templates, but its subsequent grammatical elaboration and transformation into a complete language fails to occur. This failure typically results from lack of exposure to mature speakers of the complete language. Bickerton suggests that protolanguage may provide the conceptual bridge between modern languages and the ancestral forms that the early hominids might have produced.

Bickerton identifies three types of protolanguage extant in the modern world: a) early-stage pidgin languages; b) the speech of children under two; and c) the communications of nonhuman primates taught to use signs or other symbols. All such protolanguages have the following characteristics, compared to advanced languages:

- Reduced word count per sentence or phrase.

- Key words occasionally left out, sometimes leading to unrecoverable ambiguity.

- Variable and unpredictable word order.

- Absence of complex sentence structure and elements, such as noun phrases or multiple verb clauses.

- Reduced number of inflections and purely grammatical elements such as articles and prepositions.

Protolanguage sentences often consist simply of nouns and verbs without modifiers. Speakers sometimes add common adverbial or adjectival modifiers, often attaching them permanently to particular nouns or verbs as a single rote combination. Of the three existing protolanguage groups, only normal human two-year-olds ever escape this mode of expression. Signing apes and most pidgin speakers never do, according to Bickerton.

Although rudimentary, protolanguages are generally comprehensible to advanced language speakers. Bickerton provides many examples of Hawaiian pidgin language. The following is a transcript of a native describing an object in his town. Can you tell what he means?

Building—high place—wall part—time—now-time—and
then—now temperature every time give you.

This old-timer is describing a public clock/thermometer sign mounted on the
side of a city building. Some additional examples of Hawaiian pidgin from
Bickerton:

Me no sleep. Me look. Me speak.
No can. I try hard get good ones. Before, plenty duck. Now
no more.
You no good man. You too much steal.
Me lucky. You no pi-mai (come), me ma-ke (dead).
Me no have got. Me no smoke. Me no drink. Me
Christian.

Clearly, these sentences do not qualify as proper English, but they do
represent linguistic communication. We understand the words, derive meaning
from their sequences and combinations, and form mental impressions we
believe match those of the speaker. If we could speak to these people in their
native Hawaiian, rather than pidgin, we might find them intelligent, eloquent,
and even poetic. However, without the benefit of complete English syntax,
their expressions reflect only the rudimentary linguistic structures of the
underlying universal grammar.

Bickerton's studies suggest that protolanguages, while functional as
basic communication tools, are transitory stages in the development of fully
grammatical languages in humans. Indeed, most protolanguages seldom
survive intact for more than one or two generations. Bickerton clearly
illustrates this in his studies on the children raised in pidgin-speaking
households. Remarkably, these children seldom speak the same pidgin as
their parents. Instead, they instinctively develop a more elaborate *Creole*.
Creole languages emerge when the children of pidgin-speakers reach the
critical period for language acquisition, typically at about two to four years of
age, during which most children acquire the bulk of their grammatical skills.
Instead of learning the impoverished pidgin of their parents, these children
adopt much richer forms with more elaborate syntactical structures. This
appears to happen naturally, with minimal adult coaching. Thus, in a single
generation, children naturally transform their parents' crude pidgin into their
own Creole by adding grammatical elements with greater expressive power.
A richer syntactic language appears to break through their relative linguistic
deprivation, suggesting an underlying linguistic predisposition of considerable
power.

With respect to evolutionary implications, Bickerton suggests that sign language in apes constitutes protolanguage. Research on ape signing generated great excitement when first reported (Gardner and Gardner 1977; Premack 1976; Patterson and Linden 1981; Rumbaugh 1977; Savage-Rumbaugh 1986; Savage-Rumbaugh, Shanker, and Taylor 1998). Some researchers believed that nonhuman primates might eventually communicate as equals with humans. However, while apes and monkeys can clearly acquire basic protolanguage, they seem unable to go much further. Their states of mind, as charmingly reflected in their protolanguage, seem rarely to extend much beyond their own immediate biological needs. The emergence of advanced syntactic language appears to require a much greater commitment to cognitive development than the nonhuman primates have yet achieved.

The earliest hominids might have had a capacity for protolanguage similar to that of the other apes. However, at some point, their paths diverged. Unlike the hominids, apes in their natural settings do not utilize syntax, even at the rudimentary protolanguage level, although they have the basic mental capacity for it. Some theorists even suggest that the chimpanzee vocal apparatus, although limited, may be capable of producing enough vowels and consonants to communicate verbally (Lieberman 1984; Lieberman, Crelin, and Klatt 1972; Lieberman 1991). Yet, left on their own for several million years, they never embarked on such a course. They simply did not recognize the need, nor did natural selection coerce them into it. In contrast, over the same span, the hominids developed a rich syntactical communication and used it to great effect. Thus, while the apes reach the pinnacle of their linguistic potential with protolanguage, in the hominids it represents only a starting point.

In their natural settings, nonhuman primates do communicate, just not in a recognizable syntax. Chimpanzees, for example, have some thirty-five to forty unique vocalizations, most of which appear to represent singular informational or emotional expressions (Hockett and Ascher 1964; Hauser 1996; Calvin and Bickerton 2000). Nonverbal signals often accompany these vocalizations, providing additional semantic content. Actions such as gesturing, eye contact, postural displays such as standing or walking upright, carrying sticks, or waving branches occur regularly in combination with certain vocal expressions.

Some theorists have also emphasized the importance of such nonverbal signs in the emergence of human language. Michael Corballis, for example, has hypothesized that in the early evolution of language, physical gestures may have dominated the lexicon (Corballis 2003; McNeill 1992). The development of an action-based linguistic mode could have predated the emergence of the vocal mode by several million years, potentially giving the evolving hominids an established set of syntactic templates on which the modern linguistic capability could build. This possible relationship between gestural and vocal linguistic modes has received neurophysiological support with the discovery that gestural/visual languages, such as American Sign Language, have many of the same neurophysiological and neuroanatomical substrates as verbal/auditory languages (Neville et al. 1998; Hickok, Bellugi, and Klima 1998). It appears that some of the same neural systems participate in syntactic language regardless of whether it uses the verbal/auditory mode or the gestural/visual mode.

Bickerton's studies of protolanguage give us an inkling of how our ancestors might have advanced from apelike communication to uniquely human forms. The expressive power evident in simple protolanguages, even without the embellishments of advanced syntax, suggests that the earliest languages might have conveyed a powerful selective advantage to their practitioners. The following sections outline how such linguistic forms might have originated.

Nouns and Verbs, Perceptions and Actions

One of my favorite episodes of *Star Trek: The Next Generation* has Captain Picard and a humanoid starship captain named Dathon trapped together on a deserted planet where an invisible beast threatens them. Their task, cleverly set up by Dathon as a way of making first contact, requires them to work together to defeat the alien. The catch: Dathon does not use verbs. He speaks otherwise perfect English, always a nice surprise in deep space, but his expressions consist exclusively of cryptic noun phrases referencing the mythology of his home planet. Picard, nothing if not a man of action, finds this verb-free environment frustrating, particularly as the beast intends to kill them both, verbs or no verbs. In the end, Picard overcomes his verb chauvinism and escapes. Unfortunately, Dathon does not fare as well, which may say something about natural selection for verbless communication.

I mention this episode to illustrate a fundamental point. In human experience, comprehensible sentences contain information about things (nouns) and actions (verbs). Delete either, and communication breaks down in ambiguity and confusion. Consider some of Dathon's noun phrases and their translations (Figure 4.1). Without the decoding key, his expressions mean nothing to us, however poetic or intriguing they may seem. Even with the key, we still lack a solid basis for action. Unless we understand the phrase's meaning in the alien culture's mythology, we cannot grasp its implications for action. While Dathon's crew understood him perfectly, Picard struggled. In Picard's human linguistic experience, unambiguous communication required a different structure.

Some Star Trek Phrases and Their Translations	
Phrase	**Meaning**
"Darmok and Jelad at Tenagra."	Communication (or friendship) as a result of shared struggle.
"Uzani, his army at Lashmir [with fists open; with fists closed.]"	A military strategy in which an army lures the enemy in, then attacks.
"Shaka, when the walls fell."	Failure.

Figure 4.1. Translation of some noun phrases used by the character Dathon in the *Darmok* episode of *Star Trek: The Next Generation*. Teleplay by Joe Menosky, story by Philip Lazebnik and Joe Menosky, directed by Winrich Kolbe.

It should be noted that Dathon's linguistic style is probably much closer to the norm on planet Earth (and apparently across the rest of the galaxy) than Picard's. Nearly all nonhuman species use the single-element mode of communication. When an individual monkey screams a predator warning or a bird sings a mating invitation, it is basically an ungrammatical fragment. Listeners behave appropriately only if they have the necessary preconfigured responses, that is, an instinctive decoding key. Thus, when a monkey screams "Leopard!," the other monkeys scramble into the trees. This isolated noun is sufficient because monkeys instinctively respond to it by climbing. For them, there is no need to know anything further about the leopard, such as where it is going or what it is doing. When the warning call sounds, the reflex follows. This mode of communication has obvious adaptive value and is characteristic of most communications in most species.

In their natural settings, even the great apes use this single-element mode almost exclusively. Chimpanzees rely on ungrammatical hoots and screams, augmented by gestures, postures, and facial expressions, all connected by instinctive decoding processes in the brains of the receivers. Their survival strategy demands no more.

The hominids took a different path, however. Their survival strategy relied on achieving greater control over their complex dynamic environment. Selective pressure in this arena demanded a communication mode with more flexibility. For example, in a band of ancient hominids, the distress call "Leopard!" might have prompted the same panicked scramble for the trees or rocks as we see in monkeys. However, if the caller had included an additional clarifying element in the message, the group's actions might have been different. The meaning of "Leopard Come" is quite different from "Leopard Go" or "Leopard Eat." A departing leopard gives less cause for alarm than an arriving one. A feeding leopard may actually signal an opportunity for scavenging a meal. By adding information, the sender clarifies the situation, which gives the receivers more options for productive action.

Two-element sentences, such as "Leopard Go" and "Leopard Eat," resemble Bickerton's protolanguage. One element says something about objects or things (who, what, where, when, how many, and so on), the other says something about action (movement, speed, intensity, direction, and so on). Combining such elements yields significantly more semantic precision than either element alone.

To illustrate the potential gain in expressive power, consider the following simple combinatorial exercise. Let's start with an arbitrary base of forty vocalizations, about the number of unique chimpanzee calls or English phonemes, and divide them equally into two bins, which we call *nouns* and *verbs*. We associate each noun with a perceivable object, and each verb with a definable action. Each bin therefore has twenty elements. Combining them pairwise, we can generate up to four hundred unique two-word sentences (20 x 20). Thus, even the simplest protolanguage is more expressive by an order of magnitude over our starting set of single-element messages.

Now suppose we repeat this process, but instead of the original forty vocalizations, we add the new set of two-word sentences to our starting material, making 440 starting elements in all. Each previous two-word sentence now becomes a new *two-syllable word*. For example, suppose "gah doh" originally meant "leopard eat." In the next round, the rote combination "gah-doh" comes to mean the leopard's prey animals in general, such as gazelles, bush pigs, and so on. Our original two-element sentence thus evolves into a two-syllable conceptual noun. Similarly, "hoo sik" might originally have meant "sun sleep," describing the arrival of night. The new rote combination, "hoo-sik," becomes a general indicator of action in a westerly direction, correlating to the setting sun. Here the original two-element sentence evolves into a two-syllable conceptual verb. Combining these two new compound words, we have the sentence "gah-doh hoo-sik" (leopard-eat sun-sleep) which informs listeners about certain prey animals located or moving in a westerly direction. This new construction still has some ambiguity, but it clearly illustrates the build-up of expressive power from the combination of simpler elements.

Resuming our combinatorial exercise, we follow this same logic for every original two-word sentence, and again dividing them arbitrarily into perception and action bins. We thus create a new vocabulary of two-syllable nouns and verbs, two hundred of each, plus our original one-syllable nouns and verbs, twenty of each. Combining these pairwise, we can produce 48,400 new two-word (four-syllable) sentences (220 x 220). Thus, with just a single additional combinative step, we can increase our linguistic expression by three orders of magnitude over our original forty vocalizations. Consider now that the average English speaker uses 40 phonemes to produce between 50,000 and 250,000 words (Crystal 1995). Clearly, with a relatively few iterations, even a modest set of starting vocalizations can generate enough expressive power to cover a modern lexicon. From there, the potential number of sentences grows very large indeed.

Such combinative strategies come at a price, however. Combinatorial iterations demand an exponential increase in information processing capacity. A single-element message has the mathematical properties of a simple scalar, where the two-element form is a vector. The former requires only a scalar decoding table to resolve the message, while the latter requires a vector calculus. Additional combinatorial steps produce n-dimensional vectors (tensors) which require an n-dimensional tensor calculus in the communicators' nervous systems. Each combinatorial iteration therefore requires a significant escalation of the brain's information processing capacity. Only those species that have the necessary brain mass, arranged in an appropriate architecture, can evolve from simple scalar calls to n-dimensional syntactic communication.

Nouns and Verbs in Brains

It should come as no surprise that the human brain appears to have undergone the precise physiological and anatomical transformations we might have predicted for generating the combinative gains suggested above. Moreover, these evolutionary developments are consistent with the control principles hypothesized above as the fundamental operating characteristic of functioning human brains.

The primary evidence comes from various studies using brain scan technologies. Clinical neuroscience has made extraordinary progress in the past two decades in visualizing human brain functions using various scanning techniques. Prior to the development of scanning methodologies, the study of human brain functions was limited to extrapolations either from animal models, from limited surgical interventions for therapeutic purposes, or from rare clinical cases in which individuals had survived brain injuries or neural pathologies. Obviously, animal models were of relatively little use in studying uniquely human capabilities, like language. So, for many years, the only option for studying such functions came from the rather limited base of human clinical cases.

Despite its limitations, much ingenious and productive basic research came out of the clinical domain. In particular, curious and observant clinicians formulated the first crude maps of human brain functions. With regard to language, two 19th century workers, Paul Broca and Carl Wernicke, deserve special credit. By carefully documenting linguistic deficits, known as aphasias, in various patients with known or suspected brain injuries, these researchers formulated the hypothesis that language functions depended on localized areas of the cerebral cortex. Broca's work dealt mainly with damage to specific regions of the frontal lobes, which produced an aphasia in which the patient could understand language but could not speak it. Wernicke studied patients with lesions in the margins between the parietal and temporal cortices. These individuals had difficulty comprehending language, but they could speak. The areas of the brain implicated in these syndromes still bear the names Broca and Wernicke in modern medical texts.

Starting in the 1980s, advances in brain scanning techniques gave researchers a much more dynamic and noninvasive look at human brain functions. These technologies (known by their alphabet soup designations such as CAT, MRI, fMRI, PET, SPECT, DOT, and so on) rely on the detection and visualization of markers related to the activity of local areas of brain tissue. These markers may be radioactive chemical tracers or electromagnetic emissions. When visualized, they provide a virtual map of brain activity at a particular moment in time. That map may in turn correlate with particular tasks or brain states ongoing at the time. Thus, such scans provide an indication of what areas of the brain might be involved in particular behaviors. These techniques hold huge potential for our understanding of the human brain.

In particular, scanning techniques have revolutionized our view of how the brain generates and processes human language. Specifically, it now appears that linguistic processing involves a great deal more of the cerebral cortex than the traditional Broca-Wernicke localization model suggested. Illustrative of this shift is the remarkable demonstration in many laboratories that the brain handles nouns and verbs in very different anatomical and functional locations (Pulvermüller 1999; Pulvermüller, Hare, and Hummel 2000; Damasio and Tranel 1993; Grabowski, Damasio, and Damasio 1998; Damasio et al. 1996). As illustrated in Figure 4.2, areas of the *temporal* cortex show greater activity during tests involving nouns and adjectives, while areas in the *frontal* lobes are selectively activated during tests involving verbs and spatial relationships. We have long known that frontal cortical areas have a key role in the initiation and execution of action patterns (Goldberg 2001), so it seems logical that they should also have some preferential connection to the linguistic processing of verbs. Similarly, research has implicated the temporal cortical areas in perception, conceptualization, and categorization of objects and events. Their selective involvement in the linguistic processing of nouns and adjectives again seems to make sense.

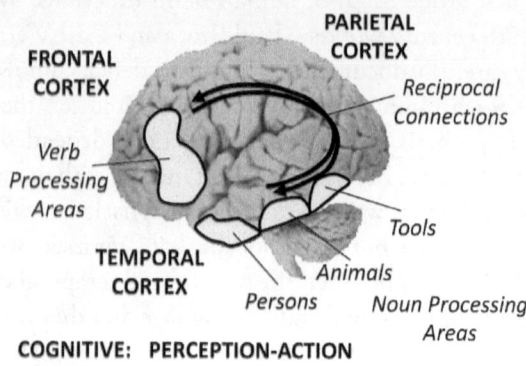

Figure 4.2. Linguistic processing areas superimposed on a photo of the human brain. Measurements of brain activity have shown that areas of the frontal cortex tend to be active during verb processing, while various temporal cortical areas are active during noun processing. Like most areas of the cerebral cortex, these areas are reciprocally connected. Such differential patterns and interactions have important behavioral, cognitive and linguistic implications.

Thus, an increasing body of research, using a variety of brain scan techniques, suggests the seemingly logical, but at the same time revolutionary, conclusion that the *same* functional brain areas that mediate high-level perceptions also process perception-related linguistic elements, like nouns. Similarly, the *same* functional areas that mediate high-level action programs also process linguistic elements related to action, like verbs. Language, rather than residing in localized language areas of the brain, as if it were some kind of independent observer or chronicler of other brain functions, instead appears intimately connected to the very processes its words and syntax seek to describe. The nouns and verbs, which *represent* perception and action, respectively, have their neural substrates in the very systems that *mediate* perception and action, respectively. This represents a fundamental shift away from the traditional model human language.

A number of researchers, notably Friedemann Pulvermüller (1999), have interpreted these new data on language processing in terms of a very old idea:

Hebbian cell assemblies, which we explored briefly in Chapter 3 of this book. Recall that Hebb's basic concept dealt with groups of neurons that become jointly active during particular brain functions, like perception, action, and so on. Because the members of any given assembly are all interconnected, the entire cell assembly can become active even when only a few of its elements become active, provided the initiating stimulation is sufficiently strong, and the other elements of the assembly are not otherwise inhibited. Hebb's hypothesis was that assemblies constituted reverberating loops whose activity might persist beyond the original stimulus. Cognitive processes reflected the sequential activation of many such loops, in Hebb's original view.

Pulvermüller suggests that the spatial overlap between functional neural processing, like perception or action generation, and linguistic neural processing, like noun or verb formation, reflect the formation of Hebbian cell assemblies that contain both functional and linguistic elements. Thus, cell assemblies that contain neurons related to perceptual processing also contain neurons specifically associated with nouns related to those same perceptions. Similarly, cell assemblies that contain neurons related to action also contain verb-related linguistic neurons. Figure 4.3 illustrates this concept. The image of, say, a leopard appearing in the visual fields generates activity in cell assemblies that contain elements related to the perception of leopards (e.g., feline shape, yellow and black color, spotted texture, etc.). In addition, according to Pulvermüller and others, those same assemblies also contain linguistic elements related to various nouns and adjectives associated with leopards (e.g., "leopard," "cat," "predator," etc.). Any of these elements, functional or linguistic, can ignite the entire "leopard" cell assembly, thereby activating all the other elements. Thus, the visual image of a leopard tends to make the word "leopard" more salient in the individual's mind. Conversely, the word "leopard" tends to make the imagined image of a leopard more salient. The complete *leopard assembly* contains *both* perceptual and linguistic information related to leopards.

The same thing holds for action concepts and words. An action cell assembly that relates to, say, climbing will contain neurons that initiate or modulate climbing movements in the motor apparatus. However, it may also contain linguistic neural elements related to climbing verbs (e.g., "climb," "grasp," "pull," and so on). As the *climbing assembly* becomes active, both the actions themselves, and the words associated with those actions, become more salient in the individual's cognitive processes. Any given element of the climbing assembly, whether linguistic or functional, may ignite the entire assembly, thereby increasing the probability both of climbing itself, and of uttering or recognizing words related to climbing.

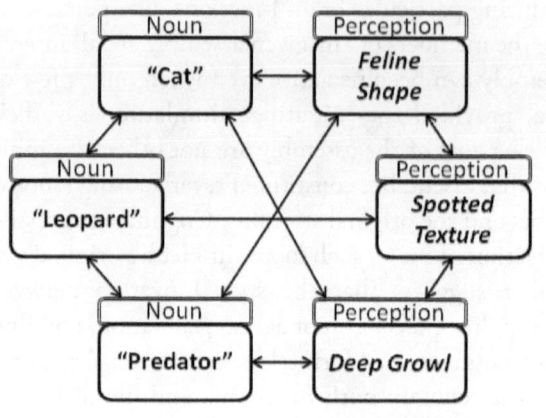

PERCEPTION-NOUN ASSEMBLY

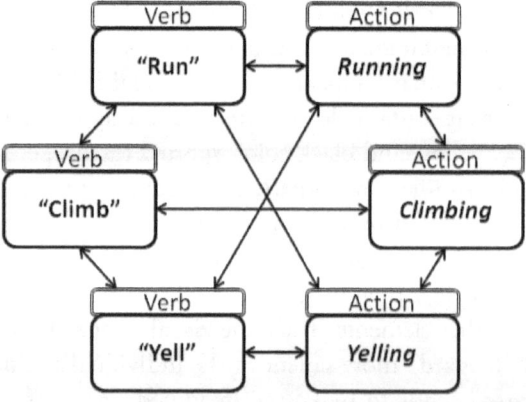

ACTION-VERB ASSEMBLY

Figure 4.3. The concept of Hebbian cell assemblies applies to the linkages between functional processes and the linguistic elements that describe them. Perceptual attributes, such as the overt characteristics of a leopard, combine in reverberating neural loops with linguistic attributes, such as the nouns and adjectives that describe leopards. Similarly, neural components of action patterns combine with verb representations that describe such actions. Activation of any one element may ignite the entire assembly. Thus, certain actions or perceptions may bias language expressions toward certain action or perception words. Conversely, activation of action or perception words may bias functional neural processes toward certain action or perceptual processing states.

The observation that the functional substrates of both action and perception also contain linguistic substrates related to action and perception changes remarkably our ideas of how language relates to other functions of the brain. Far from being a separate and localized process, as previous generations of neuroscientists believed, it now appears that language is actually an integral component of those other functions, that the *language* of perception and action emerges at the same time and in the same place as perception and action themselves. While this distributed, co-locational view of linguistic functions may challenge localization traditionalists, it is not nearly so surprising to theorists with a more dynamic outlook. If language is a tool for extracting and transmitting cognitive states, and if cognition arises from the dynamic interactions of perception and action, then it is not at all surprising that language substrates and functional substrates should be so intimately connected. From that basic realization, it is only a short step to linking linguistic functions to control architectures, as outlined in the next section.

The Language of Control

Linguistic theorists have long recognized the essential dichotomy between objects and actions. Sentences represent relationships between these elements. To be grammatical and unambiguous, a proper sentence must have at least one of each. To be informative, it must clearly indicate their interaction. All sentences in all languages that purport to describe the same interaction must ultimately resolve to the same underlying structure. More importantly, if language serves to replicate human cognition from one individual to another, it must be compatible with the structure of human cognition. This naturally involves the interplay of perceptions and actions, the primary currency of cognition, as proposed in Chapter 2 above.

This leads us to a simple hypothesis about the relationship of syntactic language and control. If cognition reflects the interplay of specific perceptions and actions, then syntactic language must comprise a symbolic extract of that basic pattern in transmissible form. The perceptions and actions that constitute cognitive processing must have their symbolic equivalents in the noun-verb combinations that constitute linguistic processing. Unlike the underlying cognitions, however, the linguistic system can release these extracts into the environment, where other cognitive entities can absorb them as new perceptions. If they have the proper configuration, these linguistic extracts, which encode the sender's cognitive state, can reconstitute that same cognitive state in the receiver, at least approximately. This reconstituted state represents a newly generated cognitive event in the receiver. Thus, where there was initially only one individual with a particular cognitive state, now there are two.

In some cases, this extraction-transmission-absorption process may reverse and the receiver now becomes a sender. He returns a linguistic extract of his new cognitive state. The original sender, who now becomes a receiver, absorbs this response, and so on. Such reciprocal exchanges constitute conversations. Repeated and embellished, they bring the conversers into cognitive synchrony, each presumably engaged in the same perception-action-expectation control process. In other words, they are thinking the same thing.

The schematic in Figure 4.4 depicts this process using the simple control schematic introduced in previous chapters. Keep in mind that this diagram describes functional concepts only, not necessarily actual neural pathways. The figure illustrates the linguistic process using the example introduced earlier of the group of hominids in need of a bridge across a stream.

As we pick up the story, the group is standing on the bank, gazing at the fruit-laden trees across the stream. We assume the group has previously seen and used natural bridges formed by fallen trees elsewhere in their range, so they should all have some sort of stored expectation of such a solution, however varied or faint. That expectation actually might have driven this group to search the local area for natural deadfalls spanning the stream, but to no avail. What this group has not yet grasped is that they can fulfill their basic expectation of a log bridge by deliberately moving logs into the stream. Absent that realization, the group remains in a state of high dissonance, as their expectation of crossing the stream on a log bridge remains mismatched with their perceived absence of such a bridge.

As the group ponders its predicament, let's focus on one bright individual who, let's say, happens to carry a somewhat stronger set of mental representations relevant to the problem at hand. Like the others, this individual scans the environment, both real and remembered, trying to make perception come into line with expectation. Like the others, this individual also twitches and fidgets, as if about to take action, but unable to initiate any particular action from the repertoire of relevant possibilities. Unlike the others, however, let's suppose this individual experiences a moment when the concept of logs, as a virtual perception, briefly interacts with the concept of pushing, as a virtual action.

The fortuitous superposition of these two concepts, logs and pushing, represents a brief cognitive interaction, that is, a fleeting thought. Now let's suppose that this interaction ignites just enough of the respective local cell assemblies to complete the internal feedback loop in the local cognitive control architecture.

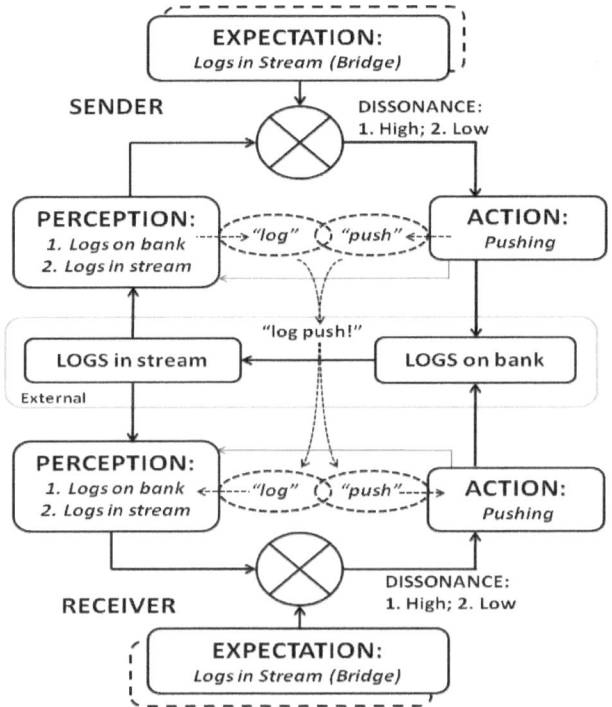

Figure 4.4. Hypothetical control interactions between a Sender and a Receiver during a linguistic exchange. A group of hominids needs to span a stream. From prior experience, all are familiar with crossing streams via log bridges (expectation) but no such facility exists in their current location. By chance, one particular Sender develops an action-perception cognitive loop consisting of Logs (perception) and Pushing (action). As this interaction develops, it creates predictive feedback that alters perception and reduces dissonance, indicating that it is a potential solution to the problem. Each element of the Sender's cognitive interaction also has word associations (see Figure 4.3) which, when expressed overtly through speech or gesture, form a syntactical expression "Log Push!" This linguistic stimulus crosses the external environment to enter the Receiver's brain, where the process reverses. Word associations activate their respective functional assemblies, which approximately reconstitute the Sender's original interaction in the cognitive control circuits of the Receiver. The Receiver may also experience a reduction in dissonance from this cognitive combination. As all members of the group develop the same cognitive state, they are all more likely to push logs into the stream. This cooperative action eventually forms a bridge, which matches the group's shared expectation and eliminates their dissonance.

As outlined in Chapter 2, such internal feedback may produce a predictive signal, that is, a momentary perceptual shift that indicates a possible outcome of a possible action. In other words, closing the feedback loop between the *log-assembly* and the *pushing-assembly* may trigger a brief imagined perception of *logs moving toward the stream*. Now, unlike all the other cognitive loops potentially swirling in this individual's mind, this one generates a perceptual match, albeit perhaps brief and partial, with the expectation of a natural *log bridge*. Even though the individual has no *actual* perception of such a log bridge, the virtual perception of logs rolling into the stream may sufficiently match the *log bridge* expectation that it produces a brief downtick in this individual's dissonance. This is the "A-ha!" moment. By chance, the control system in this one individual has produced a cognitive loop connecting the functional concepts of logs, pushing, and bridges. The momentary relaxation of dissonance stabilizes this cognitive loop as a possible solution to the problem.

Unfortunately, at that moment, the "A-ha" resides only in the mind of one insightful individual. The rest of the group remains oblivious. So what happens next? If this were a group of nonlinguistic apes, the insightful individual might attempt to act alone, running from log to log, pushing on them, gesturing wildly, and uttering frantic grunts and hoots. Others might get excited and mimic these actions until, by chance, they might move a log a fraction of the distance to the stream. More likely, however, this mob behavior would simply create confusion and unproductive activity, leaving the group exhausted and a functional bridge still nowhere in sight.

In contrast, suppose this is a band of *conlocutus* apes. Here the outcome is very different. It happens like this: As *conlocutus* individuals develop through childhood, their brains are genetically predisposed to form neural associations between the actions of their daily lives and certain sounds or gestures (verbs) that accompany them. Similarly, they instinctively associate certain perceptions with other sounds or gestures (nouns) that accompany them. This is the innate process of forming mixed functional/linguistic cell assemblies such as those hypothesized by Pulvermüller and others (see Figure 4.3). Thus, the *functional* elements depicted in Figure 4.4 (*logs* and *pushing*) also have *linguistic* elements attached to them (the words "log" and "push," respectively. As our insightful ape forms the functional cognitive interaction of *log-push*, the associated word representations in the combined cell assemblies, "log" and "push," also become active. As each becomes active, it has a higher probability of being released, that is, spoken or gestured. Since both are simultaneously active, they have a higher probability of a simultaneous release. When that happens, they *automatically* emerge in a syntactical structure: "log push" or "push log." In other words, the syntactical engine in this cognitive structure is simply the joint activation of noun and verb assemblies already tied to their respective perception and action assemblies. As indicated in Figure 4.4, this linguistic combination is simply a symbolic extract of the underlying cognitive interaction going on in the sender's mind. Its syntax (noun-verb or verb-noun) emerges as a natural by-product of the perception-action structure of the underlying cognitive interaction.

Once the sender releases them, these words pass together into the environment as auditory vibrations or visual gestures. A moment later, they enter, together, the sensory apparatus of potential receivers. Here the linguistic flow reverses. The arriving word "log" activates the receiver's cell assemblies associated with logs in the noun/perceptual parts of their brains. Simultaneously, the arriving word "push" activates assemblies in the verb/action parts. As these separate cell assemblies ignite and interact, they approximately reconstitute the same underlying cognitive interaction going on in the original sender's mind. Each receiver who hears the words "push log!" may develop a cognitive representation of *pushing logs*. As this happens, they too may experience a brief downtick in dissonance, assuming they too have some reasonably similar expectation of log bridges spanning streams. All the receivers who hear (or see) the sentence "push log" or "log push" should therefore experience the same "A-ha!" moment.

Thus, from a few sounds or gestures released by a single individual, an entire group *conlocutus* hominids replicates, however briefly and crudely, the same basic cognitive interaction as in the original sender. And like the insightful inventor, each receiver now begins to grasp the right combination of perceptions and actions required to meet their shared expectation. As the process builds, these receivers may switch to become senders, repeating and perhaps embellishing, modifying, or improving on the basic cognitive interaction. As their exchanges (conversations) continue, the group increasingly moves toward cognitive synchronization, which transforms them from a mob of frustrated free agents into a cooperative organization. As they focus their shared attention on the nearest log, and apply their combined muscular energy to the uphill side, the log rolls down the bank and splashes into the stream. They repeat the same action with additional logs, until a crude bridge begins to take shape. The match between their shared perceptions and expectations becomes progressively stronger and the dissonance felt by each member of the group becomes unequivocally weaker. Eventually, the logs span the stream. The group scrambles across, and turns to harvesting their new bounty. As they do so, these *conlocutus* hominids probably do not stop to acknowledge the miracle that just happened: that their simple sounds, arbitrary symbols communicated and recognized, gave them all a bit more control of their world.

The important point here is that words emitted by one individual triggered a cognitive synchronization in others. Simply by saying, "push log," one gifted hominid transmitted a useful solution to the group. At that moment, every member of the group was as smart as the smartest member. Their combined power brought a shared vision into reality.

Of course, this whole sequence might have happened by chance, without linguistic communication. After all, every member of the group already had some concept of crossing streams over log bridges. There were plenty of logs lying on the bank. Why could the same solution not arise by random chance? The answer is, it could. However, given the diversity, complexity, and frequency of such problems confronting the hominids in their natural environment, the linguistic strategy seems a far more effective way to transform a mob of free agents into a cooperative control organization. Random chance only works occasionally. Linguistic communication therefore afforded the *conlocutus* hominids greater relief from such vagaries.

Clearly, the value of linguistic communication is its seeming universality. As long as the relevant word associations exist, the system can generate an abstract of virtually any ongoing cognitive interaction and replicate it in a receiver's mind, with some reasonable probability of success. It happens because the structure of language mirrors the structure of cognition. The structure of cognition in turn mirrors the structure of control. Language must have its particular syntactic structure because it must flow out of and into compatible control systems. Disrupting the syntax disrupts the fit, leading to ambiguous messages, incapable of regenerating the sender's original cognitive state. Thus, syntax connects nouns and verbs in a linguistic interaction, just as control systems connect perception and action in a neural interaction. Linguistic communication has meaning only in the context of shared semantic associations, just as cooperative control has meaning only in the context of shared internal expectations.

Linguistic theorists have long suggested inextricable links between language and human cognition. The present view suggests that both are products of the fundamental control architecture that dominates the human brain. That architecture builds on recurrent loops, which, by the logic of Chapter 3, means that language also connects to our sense of self-awareness, something few introspective humans would dispute. Of course, such generalized models probably grossly understate the complexity of linguistic processes in real nervous systems. We must imagine replicating the functional elements of Figure 4.4 by the billions and interconnecting them in multiple combinations and dimensions. Sorting it out remains the daunting task facing neuroscientists and neurolinguists.

Language and Social Organizations

Cooperative control organizations represent a major innovation in the evolution of the human strategy. Coupled with our ability to predict events, as discussed in Chapter 2, our ability to organize brings the power of cooperative numbers to bear on any human problem. The emergence of language transformed individual humans into powerful social forces. These in turn propelled our ancestors out of apedom.

Humans are not unique in having organized societies, of course. Many species live in social groups, including most primates. Humans do differ in the nature of binding forces that keep their groups together, however. Robin Dunbar (1996) suggests that language plays the same role in human groups that grooming does in other primates. Grooming typically involves one individual systematically probing the fur of another, often for prolonged periods, removing parasites and other irritants. Although the action clearly has hygienic benefits, most theorists believe that it also represents a social ritual that helps bind groups together. Dunbar suggests that as the human species diverged from the great apes, conversation gradually replaced the grooming ritual. As humans evolved, they lost most of their fur, so grooming had progressively less significance as a hygienic activity. More importantly, Dunbar suggests that human social groups increased substantially in size, eventually reaching the nominal figure of about 150 individuals in a typical hunter/gatherer society. This is considerably larger than most other primate groups. Maintaining daily grooming relationships in such a large group would consume progressively greater amounts of time, leaving less for foraging, child rearing, defense, and other necessary activities. Dunbar suggests that verbal behavior replaced grooming as a more efficient means of maintaining group cohesion. Conversation allowed individuals to maintain contact with multiple social partners simultaneously, thus increasing efficiency and preserving time for other activities. In this way, conversation emerged as the preferred social binding mechanism in hominid groups.

Language-based human societies contrast with those of other primates. Chimpanzees, for example, seem the quintessential free agents, moving through their range in loosely organized bands, showing much less communication and cooperation than a comparable hominid group. Chimpanzees do have rich social interactions. However, chimpanzee groups also feature fierce competition for status and preferential access to resources. Their volatile societies can erupt instantly into a melee of screaming, running, and fighting, sometimes producing tragic consequences. It seems as if chimpanzees are less able to moderate their fiercely independent drive for dominance and personal satisfaction, often to the detriment of group cooperation. Although they have considerable intelligence, they seem unable to recognize or accept the enduring benefits of flexible cooperation, perhaps because they lack both the requisite cognitive depth and the means to communicate it effectively.

Human organizations are unique in their ability to assemble and disassemble into subgroups of all sizes, each with specific shared goals, perceptions, and behavioral repertoires. Often such groups reflect functional divisions: hunters, fishers, gatherers, builders, growers, artisans, and so on. In the course of their daily activities, groups may experience events and activities quite different from other groups. While separation could lead to loss of cohesion and fission of the group, language provides a way to restore and maintain the larger group's social integrity. Through language, separate groups can share their daily histories, so that all may recapitulate those cognitive experiences. The more vivid and articulate the language, the stronger the shared cognitions become. By continually building and reinforcing shared perceptions and expectations in the larger group, linguistic behavior allows each member to contribute to and to benefit from the experience of others.

Homo conlocutus and the Human Strategy

We started this chapter with the disturbing image of Travis Bickle in *Taxi Driver* to make the simple point that *conversation* is the crucial property of humanness, not language per se. The adaptivity of language comes from its remarkable effect on people. While the academic concepts of syntax, lexicons, and semantics may be interesting, they represent only part of the story. The other part lies in the simple biological reality that people are different after a conversation than before. The exchange of sounds and gestures reflects an exchange of minds, for better or for worse.

I chose the species name *Homo conlocutus,* the Converser, as the prototypical hominid who had the first productive conversation with a social partner. As we have seen, the structure of language reflects the control architecture resident in the cognitive mind. When we think, the cognitive interaction of perception and action entrains associated nouns and verbs. When we speak, those nouns and verbs emerge together in a natural syntax. When we listen, those words in that syntax reconstitute the speaker's cognitive interactions in our minds. Our capacity to do this is seemingly unique, certainly in terms of scale. Although other species have some of the same brain structures, they do not use them in the same way or to the same degree.

As *Homo conlocutus* evolved, the world began hearing new sounds with strange rhythms and burbling textures. The odd apelike creatures that made these sounds did not act much like apes anymore. They ranged widely in all directions, into the forests, along the streams, and out onto the plains, always with a babble of sound or a choreography of gestures punctuating their actions. At night in their camps, they faced each other and made softer sounds, not like chattering monkeys or screaming apes, but in measured phrases, interspersed with silent reflection. To an outside observer it might appear that nothing significant was happening, but every morning the group woke up smarter than they were the night before. In time, they took control of their world, working together, sharing their minds, willing and able to take on anything the world presented to them. Their camps multiplied and grew, each full of healthy children who made the same kinds of burbling sounds and stylized gestures, but steadily louder and progressively clearer.

CHAPTER 5

Homo habilis, the Technologist:
Hands, Fingers, Knees, Toes

Brain or Brawn?

In previous chapters, we have dealt primarily with events *in the head.* The transformation of the ancestral primate brain from an organ of reactive opportunism to an organ of intelligent control probably began with very little overt change in the external anatomy. These variant apes carried in their heads a richer array of reference expectations about the world, and they acted to bring the realities of the world into line with these expectations. This made them incrementally more successful than their untransformed peers, but to an outside observer they would probably have looked much like their ape cousins. Over time, however, as changes in their brain structure and function increasingly affected their development, physiology, and behavior, additional changes in their overt anatomy would have begun to accumulate. With each generation, the emerging hominid gene pool would have acquired and maintained those physical transformations that best served the changing brain and its intrinsic control functions. These changes, probably subtle and seemingly inconsequential at first, led eventually to overtly different species whose unique bones now litter various sedimentary deposits in Africa. These creatures were the ancestors that paleontologists now see as definitively hominid.

The list of anatomical changes associated with the hominid species is not overly long, but the volumes of analysis and argument among paleontologists about their relative priorities, origins, and significance are very long indeed. It seems that virtually every unique anatomical variation in the hominid skeleton, internal organs, and physiological profile has been proposed at one time or another as *the* unique characteristic and therefore as the pivotal event in human evolution. So many theories have been proposed, and ably defended by their proponents, that properly reviewing them would go far beyond the scope of this book. Instead, I shall try to build a case for one basic proposition: that all the overt anatomical and physiological variations we now see in the mainline of hominid development originated as reactions to, or consequences of, the shift in brain structure and development outlined in the previous chapters. In other words, the major physical changes *outside the head* are directly related to the changes going on *inside*. The present chapter outlines some of these key traits and leads us directly to a well-known human ancestor called *Homo habilis*, the technologist.

Bipedal Locomotion

One of the first steps on the path to *Homo habilis* was quite literally a first step, a bipedal step. Conventional wisdom puts the shift to bipedal locomotion at the start of the hominid line (Leakey and Lewin 1992; Tattersall 1998). It is one of the first definitive signs of the hominid divergence. Not that hominids have an exclusive on this ability. Nearly all primate species can walk bipedally, some with fair coordination and skill. Indeed, some authorities, most recently Aaron Filler (2007), suggest that certain anatomical adaptations consistent with bipedalism may predate the earliest known hominid fossils by some fourteen to fifteen million years. Whether any of these ancient apes actually used a ground-based bipedal gait as their main mode of locomotion remains unclear, however. Most Miocene primates were adapted primarily for arboreal life, which may have included selective adaptations for climbing, brachiation (using the arms to hang and swing from branch to branch), balancing and walking on tree limbs, and so on. Our nearest primate relatives, the chimpanzees and bonobos, retain many arboreal adaptations. They are strong climbers and spend a fair amount of their time aloft in the trees, particularly at night. On the ground, however, they tend to adopt a modified quadrupedal gait, called knuckle-walking, as do the gorillas and orangutans, with only intermittent use of the bipedal gait. Gibbons and siamangs, more distant cousins to the hominids, spend considerably more time in the trees. However, they are quite adept at bipedal locomotion over short distances when they come down.

Even though the data suggest that bipedal locomotion may have had early anatomical antecedents, only the hominids adopted it as their normal mode of locomotion. This hominid shift to obligatory bipedalism entailed substantial genetic modifications to the skeletal bones and muscles, particularly in the feet, legs, pelvis, vertebrae, and cranial attachments to the vertebral column. It also necessitated some repositioning of the internal organs and significant modifications of the nervous system, particularly in vestibular mechanisms, motor coordination, and perceptual-motor integration. As these functional changes accumulated, the hominids lost much of their facility for climbing and arboreal locomotion. Clearly, the move to permanent ground-based bipedal locomotion was an important change that did not happen overnight.

Unequivocal fossil evidence for ground-based bipedalism dates back to genus *Australopithecus*, although some hominid fossils now suggest much earlier dates (White, Suwa, and Asfaw 1994; Senut et al. 2001; Brunet et al. 2002; Filler 2007). The Australopithecines inhabited various regions of Africa starting about four million years ago. In 1974, Don Johanson and his team discovered "Lucy," the best-known member of the *afarensis* species (Johanson and Edey 1981). Lucy's nearly complete fossil skeleton has adaptations clearly consistent with those of a bipedal creature, although its skull remains apelike. The pelvis in particular appears much closer to the modern human form than any ape pelvis, either modern or ancient. That *afarensis* walked upright received dramatic confirmation by the discovery of Australopithecine footprints preserved in 3.6 million year old rock at Laetoli in Tanzania (Leakey and Hay 1979). These tracks clearly indicate an advanced bipedal walker.

Physically, *afarensis* hardly cut an impressive figure. They stood a little over four feet tall and weighed about eighty pounds, roughly the equivalent of a modern human eight-year-old. Their limb proportions, longer arms and shorter legs, were closer to those of the apes than of modern humans. Their wrist bones also retained characteristics normally associated with knuckle-walking. Their cranial capacities were roughly similar to those of modern chimpanzees (McHenry 1994; Martin 1984; Jerison 1973; Harvey and Clutton-Brock 1985), indicating that brain expansion was not yet evident when bipedalism emerged. Some researchers suggest that Australopithecine brain morphology differs from that of the apes (Holloway 1983, 1996; Holloway, Broadfield, and Yuan 2001), although this remains in dispute for the earliest forms (Falk 1983, 1992). Some paleontologists also question whether the Australopithecines are a direct human ancestor or an extinct sidetrack.

The discovery of advanced bipedalism so early in hominid evolution created a considerable stir in paleontology. To some, it raised more questions than it answered, particularly about its rather questionable selective advantages. Owen Lovejoy, for example, noted the seeming "insanity" of a bipedal gait in the open woodland that the early Australopithecines were presumed to inhabit (Lovejoy 1981; Johanson and Edey 1981). In a paradise for ambush predators and pack hunters, a little animal walking upright appears distinctly maladapted: slower running speeds, increased risk of falling, fatal vulnerability to simple leg wounds or sprains, increased exposure of vital organs, and so on. On the plus side, the extra height achieved in an upright posture might have improved the visual detection of predators. However, given the small stature of the Australopithecines in the tall grass of Africa, this hardly seems compelling.

Even worse are the implications of bipedalism for childbirth (Trevathan 1987; Rosenberg and Trevathan 1996; Tague and Lovejoy 1986). The skeletal structure necessary for efficient bipedal locomotion necessitated pelvic modifications, which narrowed the female birth canal, the passageway the fetus follows from the uterus through the vagina during birth. This opening must permit the safe passage of the baby's head and shoulders. Narrowing it increases the chances of obstetrical complications, particularly for smaller females and larger babies. Lucy's pelvis reveals a significant reduction in the size of the pelvic inlet, compared to the apes. While some analyses (Trevathan 1987) conclude that the early Australopithecines experienced only moderately increased obstetrical risk, others (Tague and Lovejoy 1986) suggest that the Australopithecine birth sequence required significant adjustments from the ape pattern. In the later Australopithecines and early genus *Homo*, obstetrical risks almost certainly increased (Trevathan 1987).

Bipedalism also complicated infant care for the early hominids. Efficient bipedal locomotion required changes in foot structure. The original grasping configuration of the arboreal ape evolved into a pad structure characteristic of a ground walker. However, eliminating the grasping hind limb created a problem for newborns. In the apes, newborns exhibit an instinctive clinging behavior in which they grasp their mother's fur using all four limbs. This serves both a protective and transport function. The loss of the grasping hindlimb would have forced the hominid infant to cling only with its forelimbs, a riskier mode. As a result, an obligatory biped would almost certainly have needed an alternative method for transporting newborn infants (Trevathan 1987).

Exacerbating these reproductive problems is the expansionary trend of the brain. Starting with the later Australopithecines, cranial fossils clearly show an expanding brain characteristic of the evolving hominids. A bigger brain means a bigger head, which requires a larger birth canal. Thus, the constriction of the pelvic inlet to support bipedal locomotion put the ancestral hominids on a collision course with their cranial destiny.

Obviously, with so many complications, the shift to bipedal locomotion must have had some major adaptive drivers. However, paleontologists have yet to reach consensus on what they are. Many theories have emerged (Kingdon 2003; Fleagle 1999; Richmond, Begun and Strait 2001). They cluster around various themes such as locomotor efficiency, tool-use, foraging strategies, thermoregulation, aquatic adaptations, sexual selection, and carrying behavior. With each new fossil find, some theories gain favor while others decline. However, no compelling argument has yet emerged as to why the hominids, and *only* the hominids, adopted the obligatory bipedal mode. Given the general capability of bipedal locomotion among other primate species, perhaps dating well back into the early Miocene, natural selection should have had plenty to work with, had there been strong selective pressure on other species, but clearly there was not. Given its obvious disadvantages, why would any sensible species follow the bipedal path?

Some researchers now believe that hominid bipedalism may not have emerged as an independent, isolated trait. Instead, they suggest, it may be part of an interrelated suite of traits that emerged together early in the hominid divergence. Bipedalism is adaptive only in the context of these other changes, but taken alone, makes little sense. This view, embraced in the present analysis, encourages the search for evolutionary connections between bipedalism and other fundamental hominid traits. Of all the possibilities, two clearly stand out. One, the remarkable human brain, we have already introduced. The other may be less familiar, but it is no less profound.

Secondary Altriciality

The evolving hominids avoided the potential catastrophic collision between pelvic narrowing and the expanding cranium by a unique developmental pattern known as *secondary altriciality*. Altriciality is a zoological term referring to the stage of fetal development at the time of birth. *Altricial* species give birth to nearly helpless offspring, lacking the motor and sensory repertoires characteristic of mature members of the species. Examples include certain birds, rodents, and most felines and canines. The opposite of altricial is *precocial*. In precocial species, newborns have more advanced motor and sensory repertoires, which enable them to perform at mature levels shortly after birth. Examples include certain bovines and equines, whose offspring can run with the herd within a few hours of birth. Together, altriciality and precociality define a continuum of neonatal development.

Primates tend toward the precocial end of the continuum. Offspring are born with eyes open and senses alert. Most newborn apes and monkeys, although helpless at birth, quickly develop sufficient coordination and muscle strength to orient and cling to their mother's fur. Within a few days, clinging enables both mother and infant to keep pace with their social group as they forage or climb into the trees to escape threats. Within a few months, most infant primates perform with a fair degree of independence.

Human offspring also display a basic precocial pattern, at least in the sensory domain. Babies are born with open eyes, they react to environmental stimuli, and they can learn and respond to parental stimulation, even in the earliest days after birth. In their motor repertoires, however, human babies appear more altricial. At birth, they are virtually helpless, unable to perform basic coordinated actions, such as clinging. They remain in this state for many days, only gradually developing the muscle strength and coordination to execute simple actions. It typically takes several months for human infants simply to turn over from back to front. This unique developmental pattern led researchers to coin the term *secondary altriciality* (Montagu 1961, 1964; Portmann 1945; Gould 1976). In human development, a secondary phase of altriciality, during which the child remains relatively helpless, overlays the normal pattern of primate precociality. This period lasts about twelve months past birth, after which the child's skeleto-muscular system grows and strengthens quickly.

Secondary altriciality is a uniquely human developmental pattern. It correlates with the evolutionary shift of metabolic resources from skeleto-muscular development to the neonatal brain. The human brain grows at fetal rates for nearly a year after birth, at the expense of body development. From birth to adulthood, the human brain increases about fourfold, the largest postnatal expansion of all primates. Brain development consumes huge amounts of energy and resources in the developing infant. During the first year of human life, the brain accounts for sixty-five percent of the total metabolic rate, compared to only nine percent for muscle tissue (Holliday 1978). The brain requires especially high amounts of fats, proteins, and other nutrients, which it consumes at the expense of other body tissues. Thus, significant postnatal brain development virtually guarantees a prolonged period of infant helplessness (Foley and Lee 1991). In other apes, such as chimpanzees, brain growth in neonates slows almost immediately after birth. Although relatively helpless at birth, chimpanzee infants develop muscle strength and coordination much more quickly than their human counterparts. From birth to adulthood, the chimpanzee's brain expands only about twofold (Gould 1977).

At first glance, the overlay of secondary altriciality on hominid infant development appears distinctly maladaptive. Increasing infant helplessness appears to be a recipe for increasing infant mortality, which, in slowly reproducing species like the great apes, should be catastrophic. How do we explain such a risky shift?

The answer lies in the advantages of building a brain to control an ever-changing world. For the moment, let's put the problem of infant mortality aside and focus on the advantages of delaying brain development into the postnatal period. This shift solves two major problems in hominid brain construction. First, by postponing development until after birth, it allows the brain to complete its development in the external environment in which it must eventually function, rather than in the isolated world of the uterus (Gould 1977). The sights, sounds, and smells of the real world, coupled with the actions of exploring, manipulating, and playing, help program the growing brain with specific information about the actual environment into which it was born, and in which it must eventually live and work. Such flexible development offers an adaptive advantage in an environment of periodic fluctuations, a condition the earliest hominids almost surely faced (Potts 1996). It clearly represents an improvement over a rigid genetic blueprint for some projected unchanging world. Late brain development potentially gave the emerging hominid species a better mechanism for coping with ecological changes through the generations.

The second major benefit of secondary altriciality is that it releases restrictions on brain size. Prior to this shift, any developmental mutation that increased brain growth faced the problem of squeezing a larger cranium through the pelvis during birth. Large-brained infants would almost certainly have had much higher mortality rates, as would their mothers. By postponing brain growth until *after* birth, however, mutations that increased brain growth would have been less risky from an obstetrical perspective. Once past the restrictions of the birth canal, the brain would be free to grow as large as biologically feasible. Only skeleto-muscular mechanics and metabolism limit its theoretical size.

Thus, secondary altriciality could have been a significant enabler both for flexible postnatal development and for increased brain capacity in the emerging hominid species. Indeed, by enabling these crucial new traits, secondary altriciality marks the start of the hominid line as clearly as any other factor. It may also help to explain obligatory bipedal locomotion, as we shall see shortly.

We do not know precisely when secondary altriciality originated in the hominid line. It is clearly a biological change of considerable complexity and evolutionary significance, which suggests that it developed over a considerable period. Theorists presume that the common ancestors of hominids and chimpanzees had the characteristic apelike developmental pattern, and that secondary altriciality emerged as a unique trait on the hominid branch. Modern apes have presumably retained something closer to the ancestral pattern. The genetic foundations of secondary altriciality will be an important topic for future research on human origins.

The Hominid Trinity

The major problem with the foregoing analysis is that we cannot put infant mortality aside, even temporarily. A larger, more flexible brain is useless if the infant dies before it matures. By all logical standards, secondary altriciality should have been a death sentence for the hominid line, an evolutionary experiment quickly extinguished by withering infant mortality. Clearly, that did not happen. The solution, I believe, was a set of compensating shifts in hominid anatomy, behavior, and social structure. One of these was the transition to obligatory bipedal locomotion.

Consider the following scenario: A female of the ancestral species common to both human and chimpanzee gives birth to a female infant. The neonate goes through a period of a few days of utter helplessness, during which the mother must carry it from place to place when she needs to move. Let's assume this species is primarily arboreal, but periodically descends to forage on the ground. Here it uses some form of quadrupedal gait interspersed with occasional bipedal locomotion. After a few days of rich breast milk, the infant gains enough muscle strength to cling to the mother's fur and support its own weight. With her infant clinging tightly, the mother can now move more freely, rejoining her social group in their daily foraging and nightly encampments in the trees. The infant accompanies the mother in all activities, learning and developing quickly. When the mother stops to forage, the infant may drop off to explore, but scrambles back to her at the first sign of trouble. When threatened by predators or marauding bands of competing apes, the mother can run or climb as necessary. If the infant loses its grip, the mother assists it with a free hand. Thus, except for the first few days of life, the infant has virtually the same mobility as any member of the group, thanks to mother's fur and a strong grip. For her part, the mother continues to participate in her social group, following the instinctive patterns laid down for millions of years.

Now let's fast-forward as that daughter ape matures, mates, and gives birth to her own infant daughter, which she dutifully cares for in the same way her mother did. Let's suppose, however, that this new infant has a genetic anomaly that delays her cortical development, extending it a bit further into the postnatal period. It might also produce a modest increase in neuronal cell count. This infant's brain completes the last stages of its development not in the isolation of the uterus, but in the external world of sights and sounds and smells. Because her developing brain receives richer stimulation from the world it will inhabit, a world of variations in climate, ecology, terrain, resources, and social interactions, this infant has the potential to mature into a smarter, more capable ape.

Unfortunately, she will also pay a price. She will be helpless not just for the normal three or four days, but, let's say, for perhaps thirteen or fourteen days. Her late developing brain draws so much metabolic energy that it retards the muscle development in her hands and feet. She cannot cling to her mother's fur.

This scenario defines, I believe, a pivotal moment in hominid evolution. The infant's extended helplessness increases the mother's care burden. Where she might have been able to stay in the same place for three to four days under normal circumstances, her increased care burden for this infant complicates her choices. If she stays in place for several weeks, the isolated pair may go hungry or fall victim to predation or attack. If she moves, she puts her infant at risk, since it cannot cling on its own. If she abandons the infant, it will certainly die, and its genetic legacy, and hers, will disappear. Thus, for her and her infant to survive during the postnatal period, the mother must carry her child as they move. This requires, I believe, a more extensive use of bipedal locomotion. This mode, already present in the primate repertoire but used sparingly, becomes the *predominant* mode during this period of extended infant helplessness.

The young mother's ability to carry her infant in this way determines the child's fate, as well as her own genetic future. If she is successful, two crucial things happen: (1) her infant survives, which means the genes for the new brain development pattern stay in the gene pool; and (2) the mother's own traits for maternal care *and* efficient carrying behavior also stay in the gene pool, inherited by her surviving daughter. In other words, the reproductive cycle in this family line contains a genetic synergy. The infant's genes, when expressed as an adult, encode *both* a new brain development pattern *and* improved locomotor mechanisms needed to transport helpless infants. With each surviving generation, this family line potentially contributes more genes for both traits. As they endure and proliferate, a subspecies may emerge with 1) a late developing brain; 2) an extended period of infant helplessness; and 3) more efficient bipedal locomotion. This is precisely the path toward genus *Homo*.

Taken together, the three major traits of hominid evolution—brain development, secondary altriciality, and obligatory bipedalism—comprise, I believe, an interrelated suite of differentiations that co-evolved early in the hominid divergence from our ape ancestors. I call this suite the *Hominid*

Trinity. Each of these traits clearly marks the path to humanness; each affects the development of the others; and, within the limits of the available data, all three could have originated early in hominid development.

Figure 5.1 illustrates the Hominid Trinity. It shows the three elements as synergistic components in a cycle. Each is both cause and effect. Thus, delayed brain development prolongs infant helplessness, which promotes bipedal locomotion to protect and care for the helpless infant. Successful infant care promotes the future reproduction of genetic variants with altered brain development, and the cycle repeats. The release of brain development from the constraints of the intrauterine environment improves the new species' adaptivity to its changing world through flexible programming, and eventually allows greatly increased brain size. Thus, we can characterize the hominid divergence as a co-evolutionary interaction among three defining features of the hominid line.

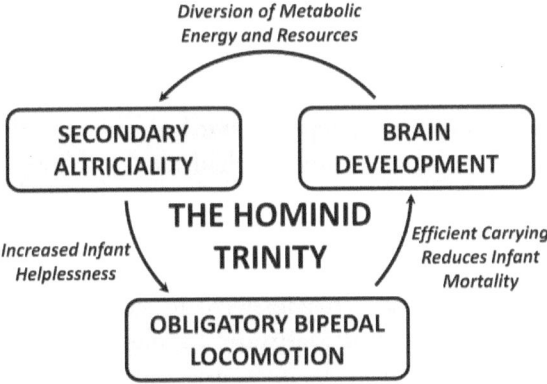

Figure 5.1. The hypothetical interplay between three fundamental hominid traits: Brain Development, Secondary Altriciality, and Obligatory Bipedal Locomotion. Starting at the upper right, a mutation in hominid brain development causes a compensating shift in postnatal development, called secondary altriciality. This allows the brain to develop in the external environment and releases restrictions on final brain size. However, this diversion of metabolic energy to the brain prolongs the period of postnatal helplessness. In response, a shift to obligatory bipedal locomotion improves parental carrying efficiency, which reduces infant mortality. The surviving infants perpetuate the genetic shifts in both brain growth and bipedal locomotion. All three traits are differentiating characteristics of the hominid divergence. Their interaction represents an evolutionary nonlinearity that could have accelerated key elements of early human evolution.

Now, some paleontologists may object to the Trinity's implied timeline. They will argue that the fossil evidence for bipedalism predates that of brain expansion by at least several million years (Walker and Shipman 1996; Leakey and Lewin 1992; Filler, 2007). The foot, leg, spine, and pelvic bones show evidence of bipedalism long before the skulls show evidence of brain expansion. How then can we assert co-evolution of these traits? The answer is that the absence of overt cranial expansion in the fossil record does not necessarily mean the absence of changes in brain development, particularly its postnatal timing and pace. The first step in making a human brain was not making it bigger. It was delaying its development further into the postnatal period. Thus, the brains of the earliest hominids could have undergone major changes in developmental timing long before any overt expansion in size. Such changes might leave no obvious fossil trace, yet still have a profound impact on the species. Paleontologists, perhaps more than anyone, understand the limitations of fossil evidence.

Logically, there must have been substantial changes in brain development in the earliest bipeds simply because, at a minimum, upright locomotion requires the augmentation of the vestibular system, and the reconfiguration of behavioral and sensory-motor repertoires to accommodate the new posture and its new perceptual and locomotor challenges. The ancestral apes had brains wired primarily for arboreal life, with occasional forays on the ground. The bipedal apes lived on the ground and walked upright almost exclusively. Their brain wiring must have changed concurrently with their bones, even though we do not see it in the cranial fossil record. Put simply, to run a bipedal machine, you have to have bipedal software. The old quadrupedal version will not do. Since the software is in the brain, there must have been continuing revisions in brain structure and function throughout the shift to obligatory bipedal locomotion. Yet the fossil record of the hominid skull shows no overt evidence of this. It is simply too crude a measurement tool to detect such subtle changes.

Understanding the interactive dynamics among bipedalism, brain development, and infant helplessness may help to clarify the earliest stages of the hominid divergence, as suggested previously by others (Amaral 2007; Etkin 1954; Iwamoto 1985; Wall-Scheffler, Geiger, and Steudel-Numbers 2007; Waters 2006). In particular, it explains why the move to obligatory bipedalism is not so "insane" after all. As maladaptive as it might seem in other contexts, bipedalism represents an obvious solution to the problem of extended infant helplessness. Such adaptations gave the helpless hominid infants their best chance to survive, which in turn allowed their new brains to deliver a better-adapted species. All of these changes, working in concert, represent a clear path toward humanness.

The Transportation Economy

Once the Hominid Trinity ignited, the new species diverged steadily from its ancestral form. Two factors shaped the next crucial phases of their external development: (1) locomotion no longer required the upper limbs; and (2) the growing brain supported a progressively larger repertoire of new behavioral patterns. Put simply, the ape had a new brain, and the new brain had hands.

While obligatory bipedalism may have originated as an instinctive response for protecting infants, it clearly had the extended benefit of enabling other carrying behaviors, such as food, sticks, rocks, and so on (Lovejoy 1981; Tanner 1981). All apes show such carrying behavior to some degree. Chimpanzees, for example, often carry food within their temporary encampments (Goodall 1986). They also pick up sticks or branches and wave them around, as either play or display. Tool-using chimpanzees sometimes carry hammer stones or twigs to a likely food supply. In some cases, female chimpanzees have been observed carrying their dead infants, a heart-wrenching image given the scenario described above.

While carrying is not uncommon among chimpanzees, it is useful only for short distances. In general, chimpanzees use their forelimbs only when seated in their typical hands-free posture: squatting on the ground with their forelimbs no longer bearing weight. What set the hominids apart was their extended use of hands and arms *during* locomotion. The bipedal gait enabled the hominids to carry things more efficiently for longer distances than other apes. Their infants and children were the first beneficiaries, but it eventually applied to all things portable. Anything they could pick up, they could transport to another location. This may seem like a trivial observation, but it marks another major turning point in human evolution.

When they started carrying things (other than their children), the hominid species transformed their traditional location-based ape economy into a transportation-based hominid economy. The ancient apes, like their modern successors, depended entirely on foraging in rich forest habitats to obtain the food, water, and the materials they needed. Remaining static for too long depleted local resources and necessitated moving on to allow the natural forest ecology to replenish the supply. Apes built their consumption economy on a location-based strategy of reactive opportunism in a setting of enormous natural wealth.

In contrast, the emerging hominids brought a rudimentary transportation system to the forest margins (Tanner 1981). This had adaptive value in their environment. They no longer needed to consume food where they found it. They could transport it from distant foraging areas to safer places for consumption. They could relocate materials and tools from their sources to their points of use. Infants, children, and pregnant females no longer needed to accompany their group on foraging expeditions. They could remain in safer locations and have food brought back to them. With efficient carrying, the hominid population gained a measure of control over the supply of certain resources. Carrying gave them a more productive way to deal with their challenging environment.

With portability came the concept of *intrinsic value*. This means simply that the utility of a thing at point *A* applies equally at point *B*. Berries that taste good on the east side of the range taste just as good on the west side, even though they might not grow there. A stone that cracks nuts at the north end of the forest also cracks them at the south end. There is no need to produce separate tools at each location. A thing's value moves as it moves. Applied across the many objects and processes of ancient hominid life, the principle of intrinsic value underlies the concept of the value-based economy, a cognitive invention that had a major impact on the human strategy (Seabright 2004).

Embedded in the concept of value-based economies is the economic principle of redistribution. With portability, goods can move from areas of abundance to areas of scarcity, enabling the continuity of supply to meet the constancy of demand. The simple act of carrying things from one place to another, replicated many times, constitutes a system of economic redistribution. To be sure, in the beginning it probably looked simply like apes carrying stuff around. But as these groups developed stable expectations around the continuous availability of food and materials, and as they measured their perceptions against those references, the act of carrying things from one place to another became a controlling economic force. In time, their foraging yields rivaled, and then exceeded those of the opportunistic apes in their richer territories.

Tools, Technologies, and Control Loops

As the transportation economy took hold, daily life for the emerging hominids began to reflect the utility of replicable *processes*. A process is simply a defined set of actions taken in sequence to accomplish an expected result. For example, a hominid cracking nuts with a hammer stone engages in a process. She places the nut on a hard surface, positions the hammer above it, pushes down sharply to crack the shell, then harvests the meat. Gathering the necessary materials and transporting them from one place to another is also a process, as is gathering the yield and distributing it to the hungry family. Processes are basically control sequences, each involving a series of unique expectations residing in the individual's brain, and a series of specific actions flowing out through the muscles. The outcomes of each action constitute a series of perceptual events, each subject to comparisons against the internal expectations. Actions may be continued or adjusted according to the outcomes of the comparisons. The entire sequence may then repeat or segue into a new process driven by a new set of expectations. As the controlling apes built their life on the forest margins, their repertoire of processes increased and diversified. In time, as their cognitive abilities grew, they added a crucial new talent, technology.

To explain technology, we must distinguish it from simple tool-use. The latter refers to a process in which the user directs force through an object or material to effect a desired movement or physical transformation of another object or material. We usually exclude the user's body (hands, feet, teeth, and so on) from this definition, leaving only noncorporeal objects and materials as true tools. Thus, cracking a nut with a rock is tool-use, but cracking it with teeth or fists is not. This may seem arbitrary, but the focus here is on objects and actions that augment an individual's genetic and physical capabilities, not on those capabilities themselves. The essence of tool-use is that it extends force and power beyond the natural limits of the limbs, organs, and tissues. Note also that the words "directed" and "desired" imply expected outcomes of these defined actions, that is, a control process.

The term *technology* encompasses tool-use but extends it in an important way. Technology is a process involving the use of tools *to make other tools*. The ancient hominids that used stones to crack nuts were tool-users; when they began smashing stones together to make sharper stones, they became technologists. Technology therefore embodies a kind of circular logic which only the emerging cognitive apes could effectively master. Technological processes typically involve sequences of tool-using behaviors, each link building on previous links, each output looping back to become an input in a subsequent step. This circular logic puts technology out of reach of less cognitive species. While simple tool-use may be well within their capability, true technology is virtually nonexistent outside the human species.

The path from basic tool-use to advanced technology follows the same control logic we have developed in previous chapters. In the context of control, tool-use is simply action taken to bring perception into line with expectation. A hungry hominid transforms hard nutshells (perception) into softer nutmeat (expectation) using a stone hammer (action). The outcomes of simple tool-use often lead directly to consummatory activity, that is, to genetically programmed behaviors such as eating, drinking, copulating, and so on. In most species, simple tool-use is tightly conditioned to such consummatory behaviors.

Consider now the crucial shift to a technological process: hammering a stone with another stone. This action also constitutes simple tool-use, but here the context is quite different. The outcome is a fractured stone, similar to the natural fragments one might gather from a rockslide. This outcome does not lead directly to consummatory activity, that is, the tool-user does not eat the fractured stones. Instead, the outcome is itself a tool, an altered version of the blunt stones the tool-user originally used. They have value only in their potential reuse as tools in some future process. Thus, the act of creating them indicates an expectation projected forward in time, that is, a prediction. In the early stages of human evolution, only the cognitive hominid, *Homo auguris* (Chapter 2), had the capacity to utilize predictive information. Only they could master the circular logic of technology: using tools to make future tools, linking sequences of tool-related actions to transform tool-related perceptions to match time-shifted expectations. With this predictive capability, the *auguris* tool-users evolved into *Homo habilis*, the technological hominid.

The first evidence of human technology came from Louis Leakey and colleagues at Olduvai Gorge (Leakey, Tobias, and Napier 1964). Dating back to about 2.5 million years, Leakey christened his hominid *Homo habilis*, the handy man. *Habilis* had similarities to the late Australopithecines and some believe they belong in the same genus. Physically, they stood about five feet tall and weighed about a hundred pounds. Their brains were considerably larger than the early Australopithecines and clearly had a more humanlike morphology.

More importantly, amidst the *habilis* bones at Olduvai, the Leakey's also uncovered an assortment of chipped stones. Dubbed Oldowan technology, the broken stones and flakes ranged from crude fragments barely distinguishable from naturally fractured rock to relatively sophisticated implements that implied an understanding of stone's potential uses. *Habilis* had clearly stepped across the technology threshold by creating cutting edges out of lumps of rock. While Louis Leakey and others focused on the *habilis* bones, Mary Leakey analyzed and categorized the tools (Leakey 1971), perhaps reasoning that if the bones answered *When*, the tools revealed *Who*.

While *Homo habilis* certainly deserves credit for mastering stone technology, the emergence of technology itself probably predates *habilis* by a good bit. Hominids were almost certainly not the first to use tools. Our ancestors probably inherited the talent from apes who used crude tools to harvest food, repel predators, and impress their mates. Unfortunately for science, the first technologists probably worked in materials that did not survive in recognizable form in the fossil record. For example, a wooden spear produced by splintering a dry sapling between two boulders might have been a triumph of ancient technology. Such a tool might have dispatched many a hare or bush pig burrowed in their earthen dens. At a modern archeological dig, however, this ancient miracle of technology would probably appear merely as fossilized plant detritus, if anything remained at all. Thus, we cannot pinpoint the moment that true technology entered the human strategy. The time gap between the ancestral apes and *Homo habilis* spans some three to four million years. The specific intermediary steps that transformed the ancient ape tool-users into habiline technologists could have begun at any point along this timeline.

Although we cannot pinpoint when hominid technology first emerged, we can be fairly certain that its progression followed the basic ratcheting principle inherent in control systems. Improving a tool-using process can mean many things: adding steps, deleting steps, changing sequences, increasing forces, decreasing forces, changing materials, altering configurations, and so on. In control terms, such events are primarily in the action domain. They represent alternative ways to transform perception to fit expectation. Such improvements often happen by chance. A random disturbance disrupts the normal flow, or an individual inadvertently changes a step, and the process generates the desired perceptual endpoint more quickly or economically. In other words, the new action represents a more effective way to resolve dissonance. Given a choice of actions for resolving dissonance, the hominid control system will nearly always default to the more efficient path. The old way simply drops out of the repertoire without a second glance. This is the technological ratchet effect.

While upward ratcheting may be a natural function of control architectures, its time course is not always smooth or predictable. A group of hominids might engage in a particular process for a considerable period, sometimes thousands of generations, without notable changes. For example, our nut-cracking hominids might have learned the technique from their parents or grandparents, who in turn learned it from theirs, and so on back for many generations. Presumably, the parents intend to pass the same skills unchanged to their offspring.

Suppose, however, a fortuitous event intervenes, say a nut-cracker accidentally discovers that sharp-edged hammer stones break nutshells with less effort, greater precision, and less waste than blunt ones. This new hammer morphology represents a potential process improvement. If the nut-crackers now search for sharp-edged stones rather than blunt ones, as their parents and grandparents did, they will realize a significant productivity gain. As a result, they may shift their search patterns to rockslides and outcrops, which feature more broken stones, as opposed to streambeds and riverbanks where traditional smooth stones are more plentiful. As they spend more time in this rougher terrain, these perceptive hominids might actually witness the formation of such sharp-edged fragments, as rocks smash together in an avalanche or rock-fall. As the expectation of sharp stones emerging from rock-falls becomes salient, these searchers may begin to trigger avalanches or rock falls deliberately to increase the likelihood of finding sharp-edged tools. Once they do this, they cross the line from gathering to manufacturing, a fundamental technological shift. These hominid tool-users are using rocks to make better rocks. In time, with the control ratchet in force, this crude process will eventually evolve into more advanced techniques for fashioning recognizable stone tools, and *Homo habilis* will emerge. Given enough time and the steady tug of the ratchet, such simple improvements will eventually take hold across the local group,

Such technological evolution requires a true cognitive brain, a proposition made evident by the relative absence of true technology among the modern apes. Chimpanzees appear to have some predisposition for technological behavior, though it is quite limited. Jane Goodall and her team at Gombe have counted nine distinct applications of tools such as stems, twigs, branches, leaves, and rocks to accomplish tasks associated with feeding, drinking, cleaning, investigating, and defense (Goodall 1986). In the best-known example, chimpanzees at Gombe insert a thin stem into a termite mound, then withdraw it with termites still attached and consume them. The chimpanzees sometimes strip leaves from a twig and shape it by chewing, suggesting an incipient technological predisposition. At nearby Mahale, chimpanzees use a similar technique to fish for ants. Interestingly, although ants and termites inhabit in both areas, the Mahale groups do not fish for termites, and the Gombe groups do not fish for ants. Further west, in Taï National Park, chimpanzees do not fish for insects at all. Here the industry is nut-cracking, an activity that neither the Gombe nor Mahale groups do. It appears that tool-use in chimpanzees has a strong cultural component, meaning that such behavior gets continually relearned by subsequent generations, passing through time as knowledge or tradition within particular groups but rarely moving between groups (Whiten et al. 1999; Whiten and Boesch 2001; McGrew 2001). Interestingly, Mercader, Panger, and Boesch (2002) have excavated sites favored by Taï chimpanzees for nut-cracking. They described the nonrandom distribution of hammer stones accumulated over many years, suggesting that this may provide a model for studying stone tool use in pre-technological hominids.

Although stone tools are the focus of much research, Tanner (1981) suggests that process improvements in carrying behavior might have been a major target of the earliest hominid technologies. Given the importance of the transportation economy, this seems a reasonable supposition. Carrying things solely with the hands and arms has limitations. The early hominid technologies might therefore have focused on simple slings, shoulder poles, or woven baskets. Such innovations would have brought greater yields with less effort, a clear economic gain.

The Hungry Brain

Technological improvements in foraging and gathering were adaptive in part because the hominids lived in a challenging habitat, but also because their growing brains represented an ever increasing metabolic challenge. Brain is metabolically expensive, consuming disproportionate amounts of energy and oxygen from the body's metabolic factory (Aiello and Wheeler 1995). Brain size correlates with the amount of metabolic energy available to sustain it (Milton 1987, 1993; Parker 1990). Leonard and Robertson (1994) found that most mammals allocate about three to four percent of resting metabolism to the brain. Anthropoid primates increase this to about eight percent, while humans devote about twenty-five percent of resting metabolism to the brain. In other words, the maintenance of human brain represents a three- to fourfold greater metabolic burden. Thus, the human energy budget differs substantially from all other animals, even our primate relatives.

Early in the evolving hominid line, the brain's increasing metabolic drain would have translated into an increasing caloric requirement. With each successive generation, therefore, a given amount of the traditional diet would have fallen progressively short, and the brainier hominids would have gone increasingly hungry. Populations that failed to respond to this challenge would have experienced chronic malnutrition with its attendant reductions in brain growth, immunological competency, and general health.

As their ancestral diet of predominantly fruits and other plant materials proved increasingly inadequate to feed their hungry brains, the emerging hominids would have to have found ways to boost their caloric intake (Leonard and Robertson 1994). The only additional source of dietary enrichment would have come from animal products: muscle, blood, marrow, and internal organs. Meat is a high caloric food, rich in protein and fat, all packed in a relatively small volume. This high yield food source clearly fit the needs of the increasingly hungry hominid population.

Meat is not an uncommon dietary supplement in primates. Nearly all species obtain some part of their calories from this source, often by eating insects or their larvae. The chimpanzee's termite-fishing behavior is one such example. More importantly, chimpanzees and other primates also consume small mammals, amphibians, birds, and reptiles. Jane Goodall documented the first clear instance of a male chimpanzee devouring a juvenile bush pig

(Goodall 1986). Initially, she could not determine whether the chimpanzees had caught and killed the animal, or had merely scavenged the meal. Subsequently, however, she observed a group of chimpanzees trap and kill a red colobus monkey that had blundered into their vicinity. The group promptly pulled the carcass apart and ate the meat. Goodall used the term "hunting" to describe this behavior. Subsequent research established that chimpanzees favor meat as a food item, but it probably does not make up more than about two percent of their overall diet.

We presume that our pre-hominid ancestors also had affinity for animal foods, although its contribution to the diet probably did not exceed that of the modern chimpanzee. In contrast, the diet of modern human hunter-gatherers shows an animal-to-plant intake ratio of about 65:35 (Eades and Eades 2001). The majority of today's hunter-gatherers get over half their subsistence from animal foods, while only about fourteen percent derive more than half their food from gathering plants. Thus, from a starting proportion of no more than a few percent, the contribution of animal products to the human diet has grown substantially. Foley and Lee (1991) estimated that a shift to ten to twenty percent animal foods would have had major evolutionary consequences. For example, a study of Australopithecine bones suggests their diet included a considerable amount of animal products, probably the meat of local grazing animals (Sponheimer and Lee-Thorp 1999). Later species of genus *Homo* probably ate meat in similar proportions, but considerably more in total volume, perhaps 35 to 55 percent more total calories than the Australopithecines (Leonard and Robertson 1994).

Increasing the consumption of animal products requires alterations in the digestive system, particularly the teeth and gut. Milton (1987) notes that hominid teeth show thinner molar enamel and considerable reductions in size compared to comparable ape teeth. Aiello and Wheeler (1995) argue that the hominid gut size varied inversely with cranial capacity, shrinking significantly as the brain expanded. A smaller gut implies a higher quality diet, richer in digestible protein and calories, yielding more energy and nutrients per unit of consumption. The hominid gut differs considerably from that of the other apes, which is larger and optimized for a diet richer in plant material. Thus, as brain activity demanded more energy, the hominid teeth, digestive system, and eating behavior kept pace, evolving in distinctly non-apelike directions.

The Technology of Meat

To our image of ancient hominid gatherers in a budding transportation economy, we must add the specter of increasing hunger. As successful as they might have been in their plant gathering activities, biochemical realities demanded a diet richer in animal products. The problem is, harvesting an animal like a wildebeest or warthog is not easy. As valuable as meat would have been, obtaining it in sufficient quantity and convenient form would not be trivial. Therefore, a major focus of emerging human technology would almost certainly have been the harvesting of meat.

Blumenschine and Cavallo (1992) have suggested that the most prudent strategy for gathering meat in great quantity, with relatively little effort or risk, would have been scavenging. While hominids probably had skill in killing small animals, they would have had less success with larger game, the richest potential source of meat. Scavenging offered a convenient alternative.

Scavenging a meal on the African forest margins involves four steps: (1) finding an animal killed or incapacitated by a predator, disease, or accident; (2) getting close to it, which may mean dealing with predators or other scavengers; (3) securing the carcass, which could include transporting it to a safer location; and (4) removing the edible portions. Although each step presents its own challenges, the last may be the hardest. Most wild animals have tough skins and tightly connected muscles. Separating the limbs and removing the meat requires either great strength, such as a group pulling together, or appropriate tools. Brute force might have succeeded for smaller carcasses, as it does for the chimpanzees, but efficient rendering of larger animal carcasses requires tools with significant mechanical advantages.

Sharp stone tools were probably the critical enabler for systematic meat processing. Once a hominid smashed a sharp stone against an animal carcass, simple physics did the rest. The path to a higher calorie diet immediately opened up. The first meat cutters probably gathered sharp stones from rockslides or outcrops. The crude edges of natural fragments might have allowed the early hominids to break through the hide, separate the limbs, and divide the meat for consumption, transport, and distribution. With sharp stone implements, even crude and unprocessed, the early hominids could have made significant breakthroughs in controlling their food supply.

As these ancient meat eaters acquired the technological insight to manufacture such sharp-edged stones, access to a richer diet quickly advanced. As scavenging techniques improved, meat contributed more to their diets, providing metabolic fuel for their developing brains to achieve their full potential. This in turn promoted more technology, as healthier brains with more neurons created the cognitive revolution.

In time, the hominids added a further process improvement to their scavenging strategy: hunting. Like their chimpanzee cousins, the early hominids probably cooperated to trap and kill smaller animals. While this might qualify as hunting, it could also be considered enhanced gathering, in the same class as fishing or digging for mollusks. True hunting began when the hominids went after bigger prey, an event of considerable significance and risk.

Consider the wildebeest. Modern animals stand about four to five feet tall at the shoulder and weigh 265–600 pounds. They can run at up to fifty miles per hour and have quick acceleration. During migrations, they can stay on the move for days. The have keen hearing, a good sense of smell, and they bolt quickly at disturbances. Even without the piercing horns and sharp hooves, bringing down such a creature poses a challenge. How would our hominid ancestors approach the problem?

Hunting has five basic elements: (1) locating signs of prey; (2) tracking to get within visual range; (3) stalking to get within striking distance; (4) striking with killing, immobilizing, or incapacitating force; and (5) extracting the usable portions of the carcass as food or materials. Except for item 4, each step seems well within the capabilities and technologies of hominid scavenger-gatherer groups. Element 4, however, is a major problem. Delivering lethal force to a large uncooperative animal demands strength, speed, courage, teamwork, and technology. Failure means not only the loss of potential food and the expenditure of energy, but also possible injury or death to the would-be predator. While the hominids might have had great courage, they had limited strength and speed. Their only hope as hunters lay in cooperation and technology.

Consider the technical challenge of applying lethal force at a distance. The sharp flakes and points produced by lithic technology could certainly produce sufficient penetrating force to kill an animal, but they would be useless if the hunter could not get past the slashing horns, hooves, and teeth to reach a vulnerable spot. Splintered wooden branches might have the requisite length, but lack reliable penetrating power. Successful hunting probably required the technological insight of combining materials and techniques, for example, using stone flakes as sharpening instruments for creating wooden pikes. Such stabbing tools combine the requisite safe length with sufficient penetrating force for at least one effective strike during a hunt. Combined with cooperative ambush tactics, such as cornering an animal in a cul-de-sac or driving it into a watering hole, a band of fearless hominids armed with sharpened wooden pikes might have had the ability to bring down a full-grown animal. Once the animal succumbed, the successful hunters could use their stone tools to cut up their kill and resharpen their weapons. On the way home, the pikes might have doubled as transportation tools, providing shoulder platforms for sharing the load.

In time, the pike would evolve into the spear and stabbing would give way to throwing, which allowed even greater safety margins. The modern human athlete can throw an object with an initial velocity of over ninety miles per hour. Allowing for the smaller frame and reduced leverage of the early humans, the best throwers might have achieved sixty to seventy miles per hour, just sufficient to overcome the quickest prey animals at close range. Such weapons gave the hominids legitimate predator status. No longer did these meat-eaters wait and watch for other predators to make their kills. They hunted for themselves. Fresh meat now contributed more predictably to their diversified economy, giving them an additional hedge against a changing environment. The growing hominid brain provided the critical competitive advantage, and now they could feed it.

Homo habilis and the Human Strategy

The transformation of the ancestral apes from arboreal foragers to bipedal technologists is truly extraordinary. Our ancestors were not the largest, the strongest, or the fastest of the apes. Indeed, they were probably the runts of the forest, pushed out to the margins where the rigors of a changing climate hit the hardest. To survive, they needed to adapt more quickly and more often than their deep forest cousins. In the initial stages of their evolution, however, the disparity loomed large between their audacious strategy of controlling the world and their limited physical attributes. The emergence of a modified brain with greater cognitive power was the fundamental response to this challenge. Its elaboration of simple tool-use into advanced technology proved the great equalizer for these runts. With their new power, they stepped forward confidently, on two legs.

Homo habilis gets credit as the first technologist principally by being the first to make tools that endured into present day. With their simple technologies, they systematically amplified the power of their puny bodies to fulfill their grander expectations. Each innovation established an expectation for the next, building a technological hierarchy to match the cognitive hierarchies growing in their minds. Each step linked perception, action, and expectation in a basic control arrangement. As genus *Homo* emerged, the recurring theme of control reverberated evermore loudly, increasingly amplified by the power of technology.

The path to technological success came through changes in both body and brain. The traditional view of technology as an outcome of bipedal locomotion is only indirectly true. A better model is that both are outcomes of a genetic shift in brain development. This change initially gave our ancestors greater flexibility to cope with environmental variations on the forest margins, but it also triggered a cascade of events that fueled the hominid divergence.

Bipedal locomotion was one outcome, not related initially to technology, but simply a necessity to protect helpless infants, the raw material of the coming cognitive revolution. As we saw above, without bipedal locomotion, infant mortality would have pushed the hominid line rapidly toward extinction, prematurely terminating the human experiment.

Once established, however, bipedal locomotion not only helped safeguard hominid infants, it helped usher in another revolution. It allowed the forest economy to shift from a location-based ape system to transportation-based hominid system. With portability, objects and materials acquired a utility that transcended location. This fundamental shift created the first value economy, a new cognitive environment in which technological improvements flourished. In this environment, hominids could move more freely, protect their young, and feed their growing brains a richer diet.

Homo habilis, the Technologist, teaches us that the same human mind that picks up simple stone flakes and sharpens wooden spears can also build machines that circle the planets and head off to the stars. Technology turns up the gain on our basic control strategy. Our puny bodies are still no match for cheetahs, elephants, or hyenas. A modern human, left on the African plain without technology, would stand little chance against them. With technology, however, humans acquire cheetah-like speed, elephant-like power, and hyena-like endurance. We can outrun, out lift, and outlast all of them, not because our muscles and bones evolved, but because our brains did.

For better or worse, modern technology reflects what the habilines and their ancestors passed down to us, how they defined us. In an age of computers and molecular engineering, their crude stone hammers and pikes seem like distant relics, but they remain deeply embedded within our psyches. If you do not believe this, then the next time you go into the wild, pick up a sharp-edged stone or a long straight staff. Feel how it fits your hand and settles neatly into a familiar cognitive niche. Walk with it awhile and realize how much stronger it makes you feel, how much more control you seem to have. Let your mind wander, and you may recognize the ancient biped who once walked the same path, and felt the same things.

CHAPTER 6

Homo bellicosus, the War-Maker: A New Recipe for Extinction

The Dark Side

The human strategy has a dark side. It is the constant threat of lethal aggression perpetrated by humans on humans, sometimes igniting into atrocity and genocide. Our facility for such behavior exceeds that of all other species. The consequences are often unspeakable: Nazis against Jews, Turks against Armenians, Bosnian Serbs against Bosnian Muslims, Hutus against Tutsis in Rwanda, the Khmer Rouge against the people of Cambodia, the list goes on. The historical record dates back at least 14,000 years to a massacre at Jebel Sahaba in Sudanese Nubia (Wendorf 1968). Some authorities suggest that it goes back much further (Ferrill 1985; Smith 2007). The circumstances of such atrocities vary, but the outcomes are eerily similar: dead humans in numbers staggeringly out of proportion to the underlying problem, at least from an outsider's perspective.

Aggression is a natural consequence of a survival strategy built on control. In the previous chapters of this book, I have presented the human strategy in a largely positive light: a plucky species, struggling for survival on the forest margins, takes a bold gamble on a brain-based behavioral shift and eventually wins the big game. However, the dark side of this story is a propensity for lethal aggression that seems inescapable. It may be the price we pay for our evolutionary success.

At its most basic, aggression is simply action taken to transform perception, that is, the outcome of a basic control process. In any complex social system, individual control agendas inevitably conflict from time to time. Two individuals want the same piece of meat, the same patch of land, the same mate. As both act to bring their perceptions into line, they may come into conflict. Each perceives the other as an obstacle to be removed. In some circumstances, lethal violence is the method of choice.

Of course, the hominids are not unique in displaying aggression within the species. Nearly all species engage in some form of social competition. The head-butting rituals of mountain sheep, the sumo-like grappling of walrus, the screaming charges of baboons, all represent forms of social aggression within the species. Even the deliberate killing of members of the same species is not unique to humans. Felines often kill the juvenile offspring of their rivals when they assert dominance over a new group. The chimpanzees, once considered the benign clowns of the forest, will murder territorial rivals under certain circumstances (Goodall 1986; Wrangham and Peterson 1996). Thus, like many other social and biological traits, the hominid propensity for violence is unique only in its frequency and intensity.

But in that frequency and intensity, human aggression surely goes beyond excess. It is one thing for social competitors to bluff and bluster, to posture and threaten, to throw objects or punches, and perhaps, in their zeal, to inflict fatal wounds on their competitors. It is quite another to use calculated lethal force to effect the systematic annihilation of other groups.

From a biological perspective, such systematic lethal aggression within the species appears seriously counterproductive. It reduces the species' reproductive base and depletes its genetic diversity, both precursors to extinction. Left unchecked, lethal aggression represents a steady push toward extinction, eventually perhaps ending the fledgling human experiment. It has not happened, or more precisely, it has not happened yet. But the past is no guarantee of the future.

If the driving theme of human evolution is the growing assertion of control over events, then we must look for links between our increased aggressiveness and our increased propensity to control. If the latter were more modest, perhaps our facility for aggression would be less intense. Conversely, as our quest for control reaches ever higher, as it surely has the potential to do, then perhaps our appetite for social aggression will grow ever larger. If control and aggression are necessarily linked, then we must ask whether human societies can ever escape their periodic lapses into organized, technological warfare with unimaginably lethal consequences. Some authorities suggest the outlook is bleak (Hedges 2002; Smith 2007). Can a species driven to control the world ever control itself?

The present chapter will address this basic question. Be warned, however. This part of the human story does not have a happy ending. We cannot walk away from our nature as controlling animals, and the drive to control, I believe, always carries the risk of excessive aggression. Absent moderation, the path of *Homo bellicosus*, as I call the first hominids to kill one another in substantial numbers, condemns us to a violent future. It is part of our nature. Fortunately, we have also acquired some moderating factors, as subsequent chapters on altruism and culture will show. There we shall find more reason for hope.

The Structure of Social Groups

Social grouping is widespread among the species of the earth. There are obvious benefits to such behavior: defense against predation, resource sharing, division of labor, technology transfer, access to reproductive partners, and so on. The hominids in particular may have benefited from strong social cooperation. Being among the slowest and weakest of the apes, and relegated to the less supportive margins of the rain forest, the ancestral hominid society may have been considerably more important for survival than that of the apes in the deep forest.

Social grouping also has obvious problems. All social partners are potentially social competitors. In principle, each member of a group competes with every other for resources and reproductive opportunities. While the benefits of cooperation tend to draw individuals together, the pressures of individual competition tend to drive them apart. This dynamic tension exists in all social groups. The crucial test of social stability is whether groups can moderate competitive disruption in order to optimize cooperative benefit. Some succeed, and move forward in time. Others fail, and slide quickly and quietly into extinction.

The study of social systems has been a mainstay of psychology and sociology for many decades. Researchers have traditionally focused on structural factors, such as group composition and size, shared values and qualities, rights and priorities, roles and functions, and authority for decision-making (Goldschmidt 1960). As a species with powerful cognitive capacities, human societies incorporate some additional abstract dimensions, such as culture and ideology, which help stabilize and codify the group's social rules and traditions. Understanding social systems requires understanding such structural factors and how they interact. When social systems break down, as they sometimes do, social theorists invariably look for deficiencies or unresolved issues in the fundamental elements of social structure.

The addition of evolutionary biology to the traditional psychological and sociological approaches creates the discipline of *sociobiology*, which treats social systems as evolving biological entities built on genetic foundations.

Sociobiologists assert that social behavior, like all biological phenomena, must have adaptive value in order to perpetuate over time. If certain practices, such as excessive aggression, prevent a society from reliably producing offspring, it must eventually go extinct. Conversely, societies so passive or weak that they cannot defend their territory must eventually give way to more aggressive groups. Social systems must find a stable balance between competition and cooperation in order to endure over time. Sociobiology strives to characterize such balances and to explain how particular species and groups achieve and maintain them.

Taking a strict evolutionary view, sociobiology emphasizes the interplay of a species' ecological challenges and its genetic make-up in determining the adaptiveness of its social systems. One cannot simply assert that aggression is always maladaptive, or that peace and harmony are always adaptive, without understanding their dynamic roles in the social system. Moreover, one also needs to understand how that social system copes with its environment. Aggression may be entirely appropriate in one social system, but fatally disruptive in another. The biological measure of social stability is endurance through time, not some ideal of social harmony or personal fulfillment. Groups that display persistent discord and aggression may endure, while seemingly utopian societies may succumb in just a few generations.

Human society is but one of a wide variety social systems across the biological spectrum. In some species, social structure is relatively minimal, consisting of little more than periodic mating, followed by long intervals of solitude. In others, like human beings, they are permanent entities encompassing large numbers of individuals over vast geographic territories. Obviously, the wide disparity in social behavior implies a wide diversity in evolutionary history and selective pressures. Even among close genetic relatives, such as the humans, chimpanzees, and bonobos, there are major social differences. Such strong variations among genetic cousins complicate models of social evolution (de Waal 2001).

To facilitate the discussion of social structures, I shall divide the universe of social interactions into five basic categories:

- Sexual bonding

- Kinship

- Inlawship

- Friendship

- Xenophobia

All social systems incorporate these elements in greater or lesser degrees. Each element contributes to the overall balance of cooperation and competition, as shown in Figure 6.1 and discussed further below. Together, these elements define the root social structure of a particular species, including its overall propensity for aggression.

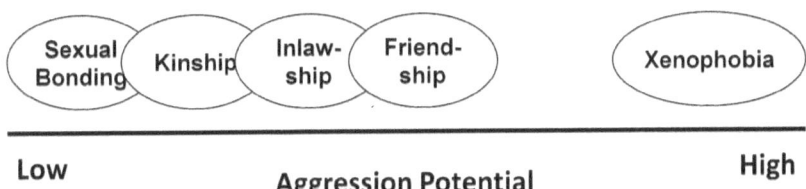

Figure 6.1. Human social components mapped on the dimension of Aggression Potential. Sexual interactions represent the least aggressive social system. Neither partner gains by aggression if it compromises the other's reproductive fitness and disadvantages their shared offspring. Kinship systems have relatively low aggression potential depending on their members' degrees of relatedness. Inlaw and friendship systems have more potential for aggression, as the benefits of cooperation are more indirect. Xenophobia defines the extreme end of the aggression spectrum, reflecting its role as a major causal agent for warfare and lethal violence.

Sexual Interactions

In evolutionary biology, reproductive fitness is the ultimate validator of all behavior. It is the nexus through which all lines of biological inquiry must pass. For most of earth's complex multicellular species, sexual reproduction represents the primary mode. Its fundamental purpose, the fusion of gametes from two separate individuals to produce progeny, defines a root level of social cooperation, the prototypical social system. Whatever may happen before or after, sexual interaction constitutes a core social behavior on which all other social systems must build.

As a social interaction, sexual behavior has unique properties. First, as an evolutionary trait, it is by definition sexually selected. Darwin and his contemporaries described sexual selection as the process by which a species disproportionately retains phenotypic traits that promote successful sexual reproduction. Thus, certain physical attributes, behaviors, and preferences may emerge simply because they succeed sexually. The classic example is the outlandish plumage of the male peacock. The progeny of such matings express the same traits and preferences when they mature. They become characteristics of the species. Conversely, traits that reduce sexual success disappear quickly.

Second, sexual behavior is subject to genetic variation, as are all biological processes according to evolutionary theory. The social patterns of sexual behavior may vary from one generation to the next because of the mutation and recombination of genes. Unlike other genetic changes, however, variations in sexual behavior are subject to nearly instant selective feedback in the individual's reproductive success or lack thereof.

Third, sexual preferences represent one of the defining characteristics of species. Biologists characterize speciation, the emergence of new species, in part as the formation of exclusive reproducing populations. Members of an emerging species tend to mate within their groups but not outside, even though they may be genetically compatible with the larger population. Reproductive exclusivity may result from geographic isolation, variations in physical attributes, or behavioral preferences. Each of these factors may influence sexual interactions and thereby drive speciation.

Fourth, reproductive social behavior has multiple phases, each of which may show distinct variations among species (Morris, 1967). For convenience, I define four categories: (a) prenuptial signaling; (b) acquiescence or bonding; (c) consummation; and (d) post-fertilization activity.

Prenuptial signaling consists of species-specific behavioral displays, sometimes accompanied by physical and/or biochemical changes, which indicate sexual fitness and readiness. Species have evolved an enormous variety of prenuptial signals. Human forms include postural, linguistic, and contextual cues, coupled with cultural augmentations such as clothing, adornments, and scents. Also included, but often overlooked, are subtle physiological or biochemical indicators involving the olfactory, tactile, and gustatory senses.

The *acquiescence* or *bonding* stage of sexual behavior varies from the simple invitation to copulate, indicated by postural changes or gestures, to the formation of complex relationships that include permanent biochemical and neural changes. The human species tends toward the latter. A courting human couple may form a pair-bond, a relatively long-lasting relationship that appears to have both behavioral and physiological foundations (Carter and Keverne 2002; Insel and Young 2001). Some theorists relate it to the ethological phenomenon of imprinting (Morris 1967; Lampert 1997), while others suggest a considerably more complicated process (Symons 1979). While cultural traditions have mythologized the human pair-bond into an ideal that biological reality seldom achieves, pair-bonding does represent a unique pattern that differentiates our reproductive social system from others, including the other great apes (Diamond 1992).

Consummation includes all the behaviors and physiological reflexes involved in accomplishing fertilization. This typically involves the ejaculation of sperm by the male into the vicinity of the female's ova. As the proximal event in reproductive success, consummatory behavior is subject to strong sexual selection and is largely instinctive and reflexive. In mammals, fertilization initiates a series of physiological, developmental, and behavioral changes associated with pregnancy and fetal gestation.

Post-fertilization activity includes behavior associated with pregnancy and preparation for childbirth. In mammals, females carry the major burden in this area, while males vary greatly in their levels of support, depending on the species. In chimpanzees, males seldom recognize or acknowledge pregnancies for which they are responsible, leaving females to carry the entire burden. In contrast, human males often maintain contact with pregnant females, offering protection and support as part of their pair-bonding arrangement. The birth of offspring completes the post-fertilization phase and initiates a new class of social interaction around parenting and kinship, as outlined below.

Species may vary significantly in any or all of these sexual/ reproductive elements. Even closely related species, such as humans and chimpanzees, show wide variations. For example, chimpanzees and humans differ substantially in both the bonding and post-fertilization phases. In chimpanzees, estrous females will acquiesce to sexual intercourse with almost any eligible male in the group, showing little, if any, pair-bond commitment. Post-fertilization activity by the successful chimpanzee male is almost nonexistent. He provides little or no direct support to the gestating female, other than participating in the general group defense against predators or outside groups. Humans and chimpanzees display strong social differences in these areas, even though genetically we are closely related species.

Of all the social interactions, sexual behavior is the most cooperative and least competitive. The product of sexual interaction, offspring, necessarily reflects equal genetic contributions of both partners. With respect to reproductive fitness, both parties benefit from cooperation and neither gains much from direct competition. Aggression between them has no benefit at all if one partner kills or injures the other before achieving the desired reproductive outcome. Sexual interaction is therefore inherently less competitive and more cooperative than other social interactions. Figure 6.1 therefore depicts sexual behavior at the extreme low end on the axis of social competition.

Kinship Interactions

Successful sexual interactions produce offspring. Their arrival defines the class of *kinship* social interactions, which encompass unique behavioral repertoires among groups of individuals who share a portion of their genes. The prototypical kinship interaction is the parent-child relationship, but kinship groups may include all genetically related individuals such as siblings, cousins, aunts and uncles, grandparents, and so on.

Parental care varies greatly among species. In some, neither parent participates. In many fish, for example, the male and female interact sexually to produce fertile eggs, then both disappear, leaving the eggs to hatch and the offspring to start life alone and unaided. In most mammals, at least one parent remains with the offspring for some period of time, providing food, warmth, cleaning, protection, stimulation, and/or instruction. In primates, the female generally takes the primary role in this area. Males vary considerably in their involvement. Chimpanzee males remain almost entirely uninvolved, while human males often provide almost continual parental support.

Like sexual behavior, parental care is subject to strong sexual selection. Successful parenting consolidates reproductive success by preserving and transforming offspring into sexually viable adults. Theorists have developed a classification scheme for the various patterns of parental investment (Pianka 1970). They designate species as *r-selected* if they have minimal parental investment, but higher overall numerical production of progeny per parental mating. Such species rely primarily on enhanced fecundity to increase the chances of sustaining their genetic lines. The opposite, *K-selection*, refers to species that emphasize the survival of each offspring through greater parental investment. Such species typically produce fewer progeny, which have longer gestation times, extended infancy, prolonged childhood, and many other physical and behavioral differences (Wilson 1975; Hrdy 1981; Foley 1987).

Primates tend toward the K-selection pattern, with humans showing a particularly strong form. Indeed, the phenomenon of secondary altriciality (see Chapter 5) puts humans in a unique position among K-selected species. The human infant is so helpless for so long that successful reproduction requires a considerable increase in parental investment, compared to other primate species. Unlike the chimpanzee, where the nursing female can travel and forage relatively unencumbered, a human female in a comparable environment would be hard-pressed to care for both herself and her helpless infant. Alone and unsupported, even with powerful maternal instincts, she has less chance of success. If her infant dies, both she and her male partner share the reproductive failure.

Arguably, secondary altriciality affects all aspects of human parenting behavior. For example, our propensity for pair-bonding is unique among the apes (Lampert 1997; Fisher 1992; Diamond 1992). While it does not guarantee higher birth rates, it does promote higher neonatal survival rates. The male's continuous support of both mother and infant improves the overall likelihood of reproductive success. Indeed, so valuable is joint parental investment that human sexual behavior, particularly in the prenuptial stage, may include assessments of parenting potential. Females obviously have a higher concern for the male's parental intentions, since females are most vulnerable to abandonment. However, human males also have a stake in selecting females who project strong attributes for extended maternal care.

Theorists have also argued that the evolution of human gender patterns, both physical and behavioral, reflect the unique needs of human parenting (Geary 1998; Wilson 1975; Trivers 1972; Daly and Wilson 1978; Alexander and Noonan 1979). The compelling evolutionary pressure against male abandonment has produced anatomical and physiological changes in the female such as continuous sexual receptivity, concealed ovulation, permanently enlarged breasts, and menstrual synchronization. These traits conspire to maximize male interest and constancy, which serves the female's strong K-selection strategy (Symons 1979; Benshoof and Thornhill 1979). Human males have also undergone significant evolutionary changes, for example, reduced sexual dimorphism and a stronger predisposition for pair-bond arrangements, as compared to other male apes (Diamond 1992).

The birth of children to a pair-bonded couple creates a nuclear family, the cornerstone of kinship groups. The nuclear family encompasses both parent-child and sibling relationships. In time, as the offspring mature and reproduce, the kinship system tends to expand to include grandparents, aunts, uncles, nephews, nieces, and cousins. Kinship groups typically involve many individuals in complex patterns of relatedness. Two principles govern kinship social systems: 1) inclusive fitness; and 2) genetic diversity. Their dynamic interactions produce a variety of interesting social phenomena.

The theory of inclusive fitness, also called *kin selection*, says that a biological entity can accomplish reproductive success either by passing genes directly to progeny, or by ensuring the reproductive success of relatives who share the same genes (Hamilton 1964; Williams 1975; Dawkins 1976). Under this theory, members of a kinship social system have a stake both in their own personal reproduction, *and* in the reproductive success of closely related family members.

Richard Dawkins (1976) offered a unique perspective on this problem by invoking the *gene's eye view*. The driving force in kinship groups can be represented logically as a gene's attempt to make more copies of itself (the *selfish* gene view). A given gene might do this by promoting behavior that leads to its own replication, or by promoting behavior that leads to replication of copies of itself residing in other members of the kinship group. From the gene's perspective, these are equivalent. Extensive analysis and experimentation suggest that real kinship social systems obey this genetic logic. Group members strive both to reproduce themselves and to promote reproduction by others in direct proportion to their genetic relatedness.

Among closely related kin, such as parent-child or siblings, cooperation dominates the social repertoire. Kinship preferences underlie food-sharing, defense against predators, and reproductive opportunities. Such mutual support maximizes inclusive fitness by insuring the enhanced survival of shared genes. Although siblings may sometimes compete for parental attention and resources, direct aggression is usually rare, in accordance with the principle of inclusive fitness.

While kinship selection is a powerful determinant of human social behavior, it cannot be the only governing factor. If it were, the ideal human reproductive strategy would be incestuous mating among the members of the nuclear family: male siblings should pair with female, parents with children, and so on. This represents the optimal way to promote shared genes into future generations. In actuality, of course, this rarely happens. Most species, including humans, avoid such incestuous pairings through innate biological mechanisms (Shepher 1971; Wolf 1968, 1995; Fox 1980) and strong cultural traditions. The primary reason for this is that incestuous reproduction violates the second governing principle of kinship groups, *genetic diversity*.

Genetic diversity is a potent, but often overlooked, force in social evolution. It reflects the fundamental requirement of evolving systems for genetic variation. Expressed variation is the substrate of natural selection. Without it, systems cannot evolve, and without evolution, they cannot sustain themselves in a variable environment. All evolving biological systems *must* incorporate mechanisms to insure the introduction, accumulation, and expression of genetic variability in the population. Incestuous reproduction violates this rule. Prolonged inbreeding reduces diversity and, carried to the extreme, promotes genetic uniformity. In artificial situations, such as laboratory animals or agricultural products, this may be desirable. However, in natural settings with dynamic climates, resources, predators, and pathogens, genetic uniformity is a recipe for extinction. Inbred species are inherently vulnerable to change while diverse populations have a greater chance of perpetuating the species (Ridley 1993).

In social systems, the biological drive for genetic diversity promotes strong reproductive preferences for individuals outside the recognized kinship group. Members of one group tend instinctively to pair with members of other groups.

The genetic basis for such preferences remains a topic of active research. Such exogamous behavior patterns provide a powerful counterbalancing force to the selfish gene's promotion of pure kin selection and its attendant temptation to incest. At first glance, the underlying pressure for genetic diversification appears to weaken the power of kin selection. However, in the long-term, that is, over evolutionary time, it actually protects the kinship group against extreme genetic uniformity and its attendant risk of extinction.

In Figure 6.1, kinship interactions occupy the lower end of the aggression spectrum, indicating their relatively strong tendency for cooperation and avoidance of severe competition. Such interactions are somewhat less cooperative than sexual interactions, however. The frequent squabbling of siblings, and occasional feuding of various family members, reflect greater aggressive potential in kinship groups, as compared to pair-bonds. Indeed, kinship social structures can sometimes become catastrophically dysfunctional, leading to tragic acts of aggression.

Inlawship and Friendship Interactions

Like kinship groups, inlawship social systems also provide stable frameworks for assuring genetic diversification while maintaining the principles of kin selection. However, there are some significant differences. Inlaw relationships develop when members of two different kinship groups mate and produce offspring. Such pairings obviously satisfy the principle of genetic diversification, but in the process, they also produce children, in whom *both* kinship groups have a shared genetic interest. Clearly, this also invokes the principle of inclusive fitness. Inlaw social systems thus constitute a form of *indirect kinship*, in which all parties gain directly by supporting their related offspring and indirectly by supporting each other. Thus, inlaw systems represent extended cooperative arrangements among otherwise unrelated individuals.

Friendship represents a similar affiliation of unrelated individuals held together by a kind of *surrogate kinship*. If individuals *A* and *B* are unrelated, but *A* consistently acts to support *B*'s genetic fitness, then it may be in *B*'s interest to support *A*'s fitness as well. This defines a friendship system. The inclusive fitness conveyed by such a system is indirect, as the parties do not necessarily share significant genetic kinship. However, friendship systems may represent a preliminary stage in the formation of inlaw relationships, as friendship circles provide a readily accessible pool of reproductive partners.

While inlaws and friends may be predisposed to cooperate, they also have the potential to compete. The bonds of inclusive fitness in such groups are entirely indirect and therefore inherently weaker than in true kinship groups. If tangible benefits do not accrue from this cooperative social arrangement, the participants may see competition as a more satisfactory course. Therefore, inlawship/friendship groups are inherently more volatile than true kinship circles. Unless the indirect bonds are very strong, aggressive competition can disrupt the social structure. Figure 6.1 therefore places inlawship and friendship in neutral positions on the axis of aggression potential.

Combining and Optimizing: SKIF Communities

In real social systems, the four social interactions described above, sexual bonding, kinship, inlawship, and friendship, typically operate together to serve the goal of species reproduction. Each contributes in greater or lesser degree, depending on the composition of the group. Together, these four elements define an aggregate social entity, which I call the sexual/kinship/inlaw/friendship (SKIF) community. SKIFs represent the stable core of most human societies, as well as many other primate species (Dunbar 1996). Each of the four major social components contributes to the unique balance of social cooperation and competition in the SKIF. As an aggregate social structure, SKIFs are inherently less cooperative than pure kinship or pair-bond social systems, but more cooperative than a purely random assortment of unrelated individuals.

Depending on circumstances and composition, a given SKIF may endure for extended periods as a stable social cooperative. Alternatively, it may become unstable and fracture into competitive, sometimes hostile, subgroups. Stability depends on the precise titration of cooperation and competition within the SKIF community. This is a fundamental problem of social optimization, which all enduring social systems must solve. Several well-known mechanisms contribute to this optimization process.

Perhaps the most familiar is the *dominance-subordinance hierarchy*, one of the classic concepts of social theory. Most students are familiar with dominance-subordinance hierarchies as the classic *pecking orders* in barnyard fowl (Schjelderup-Ebbe 1935). This avian model has influenced social theorists for decades. However, research has shown that in most species, particularly primates, the theory of dominance-subordinance is considerably more complicated. While it often focuses on competition, with an emphasis on its aggressive and occasionally violent manifestations (Schjelderup-Ebbe 1935; Lorenz 1963; Passingham 1982; Dunbar 1988; Wrangham 1987), dominance-subordinance hierarchies are actually elaborate social optimizing mechanisms for balancing cooperation and competition within a community. The resulting social hierarchies reflect the underlying dynamic interplay of inclusive fitness and genetic diversity.

Social scientists and ethologists traditionally define dominance as a relationship between two group members in which one has favored access to some object or situation to which both aspire. Desired goals include almost anything that conspecifics might value: food, shelter, territory, mates, deference, acceptance, protection, and so on. In its classic form, social dominance reflects the struggle for reproductive rights. For example, suppose a single female reaches reproductive maturity and two bachelor males decide to pursue a sexual liaison with her. The situation is inherently competitive as only one male may impregnate the female at a given time. Obviously, with reproduction as the prize, the outcome is not biologically trivial. The genes of winners may endure while those of losers may disappear. According to classic Darwinian Theory, dominance behavior should therefore be subject to strong sexual selection. In many species, such sexual dominance derives from physical attributes such as size, strength, and aggressiveness. When sexual competition escalates to physical fighting, the more physically intimidating individual wins by instilling fear or inflicting punishment on the other. Subsequent reproductive success then perpetuates these winning attributes into the following generations.

In rare instances, dominance struggles can escalate into lethal violence in which the stronger kills the weaker. By killing rivals, dominant individuals guarantee exclusive reproductive rights, passing on the same aggressive traits to their progeny. Continued to its logical extreme, such a species should eventually evolve traits of extreme physical development in a social system in which unchecked lethal violence is the norm among sexual rivals.

Such violent societies are rare, however, as moderating forces invariably intervene. The primary evolutionary restraint on runaway dominance is the necessity for genetic diversity. If only a few dominant individuals ever reproduce, the species necessarily regresses toward inbred uniformity. Those few account for a steadily increasing proportion of the gene pool, putting the species at increased risk of extinction. Clearly, there must be biological mechanisms that prevent such runaway dominance from jeopardizing genetic diversity. Several such *egalitarian* mechanisms exist. For example, kinship itself. If potential competitors share kinship, then lethal aggression between them is genetically counterproductive. The death of either reduces their shared gene pool, which reduces their inclusive fitness. Indeed, kinship partners often cooperate to maintain or improve jointly held dominance positions against other members (Hausfater, Altmann, and Altmann 1982).

Incest avoidance also moderates runaway dominance by promoting reproduction outside the dominant line. If they are to avoid inbreeding, the offspring of dominant individuals must avoid one another. In seeking outside mates, they necessarily disperse the dominance genes, effectively diluting the gene pool that might otherwise produce runaway dominance behavior.

In the human species, pair-bonding has an additional and unique moderating impact on dominance competition. Implemented fairly, pair-bonding effectively guarantees reproductive access for more group members, as it tends to remove the previous winners from the competition, leaving fewer rivals in future tests. The losers in one competition may fare better in the next, since they need not face the winner again. Compared to a non-pair-bonding species, such as the chimpanzee, pair-bonding tends to moderate the frequency and intensity of reproductive dominance conflicts, theoretically producing a more harmonious and cooperative social group. Of course, pair-bonding is not foolproof, and philandering can trigger intense conflicts in human groups, sometimes resulting in lethal violence.

In most species, dominance-subordinance competitions have evolved into behavioral rituals that rarely cause injury or death to the parties. Competitors engage in elaborate displays, bluffs, and vocalizations intended to intimidate rivals, but they seldom escalate to actual violence. If the display is successful, subordinates display characteristic submissive behaviors that end the competition. The ritual outcome defines each individual's place in the social system. Clearly, nonlethal competition has advantages in preserving genetic diversity, while allowing fitter members to acquire higher status in the group and thereby take a larger share of its resources.

Field research has documented several interesting examples of egalitarian moderation in primate dominance hierarchies. In several species of monkey, for example, researchers have found that lower ranking members, as a group, sire about as many offspring as the higher ranks (Bernstein 1981; Dewsbury 1982; Ellis 1995; Altmann et al. 1996; Pusey, Williams, and Goodall 1997). Such couplings often occur covertly, out of sight of the dominant members. Even more remarkable, studies in macaques (Berard et al. 1993, 1994) showed that higher-ranking individuals sometimes abandon their preferred position and emigrate to other troops, where they assume a lower rank. In part, this occurs because females in their original troop apparently lose interest in familiar males, even those of high rank. In chimpanzees, dominant males tolerate reproductive behavior by subordinate males as a coalition-building social strategy (Goodall 1986). Chimpanzees outside the coalition must resort to covert sex. In a large group, spread out over a wide geographic area, the higher ranks apparently cannot prevent sex in the lower ranks, which insures the continual diversification of the gene pool.

While so-called dominance hierarchies clearly have a role in primate social structures, they rarely reflect the rigid *pecking order*s of classic social science. Instead, they represent a more fluid balancing mechanism that rewards individuals with certain traits a larger share of resources and preferential reproductive access, but without sacrificing the overall genetic health of the species. While individual dominance struggles always have the potential to escalate into lethal aggression, which may reward the victor with a short-term gain, its extension into a runaway aggressive strategy represents an evolutionary dead end.

All human communities maintain a complex dominance-subordinance structure. Within the complex mix of pair-bonds, kinships, inlawships, and friendships, certain individuals achieve preferred access to resources and reproductive partners. Certain physical attributes may promote this preference, as well as other traits and talents, such as behavioral, economic, or cognitive. Although dominance struggles may sometimes erupt violently, human SKIFs tend to be relatively egalitarian, cooperative, and supportive, at least compared to other primate species. If all human social interactions remained entirely within the local SKIF structure, human life would be relatively free of aggression and violence. Unfortunately, outside the relative warmth and safety of the SKIF community, the situation changes drastically.

Xenophobic Interactions

The relative harmony within SKIF communities stops at the boundaries between rival SKIF communities. The members of one SKIF recognize and accept one another, but they typically treat outsiders with wariness, suspicion, and hostility. Even though they are members of the same species, with the same general attributes, behavior, and genetic potential, their social interactions reflect strong *xenophobia*, the primary social element that leads to conflict, violence, and atrocity. As depicted in Figure 6.1, xenophobic interactions are at the high end of the aggression axis.

Social theorists distinguish between the ritual dominance-subordinance aggression practiced within SKIF groups and the xenophobic aggression directed toward outside groups (Cheney 1987; Manson and Wrangam 1991; Tooby and Cosmides 1988; van der Dennen 1995; Wrangham and Peterson 1996). While xenophobic aggression has some of the same characteristics as dominance-subordinance aggression, it has fewer and weaker moderating influences, which tends to produce more frequent and more violent conflicts.

The contrasting dynamics of within-group *versus* between-group social competitions have become major determinants of human social evolution.

Xenophobia reflects two synergistic forces. The first is the relative absence of perceived kinship between SKIF groups. By definition, separate SKIF groups represent separate gene pools, at least to a first approximation. The cooperative forces of pair-bonding, kinship, inlawship, and friendship are weaker between SKIFs, which diminishes their aggression-moderating effects. Reproductive exclusivity also tends to promote the perception that outside groups are somehow outside one's own species.

The second driver of xenophobic aggression is resource competition caused primarily by population pressure (Lumsden and Wilson 1983). Successful species tend to increase their populations, which necessarily creates problems. A fixed supply of any resource yields progressively less per capita as the population grows. To sustain itself, the population must either expand its resource base, which increases competition among neighboring groups, or accept deprivation. Deprivation, however, creates internal pressure within the SKIF social structure, leading to potential fractures and upheavals. Expanding the resource base is the only solution to this problem. Thus, any imbalance between supply and demand, such as might occur in a period of climatic variation, must lead to resource competition between groups.

The chimpanzees of Gombe represent a case study of the potential consequences. After observing chimpanzee populations living relatively peacefully for many years, Jane Goodall and her colleagues watched in shock as *war* broke out between two neighboring groups in early 1974 (Goodall 1986; Wrangham and Peterson 1996). The initial attack involved a group of eight members of the Kasekela group who entered the home range of the neighboring Kahama group and killed a lone male. During the following months, lethal raiding continued until nothing remained of the Kahama group. Quite simply, one group of chimpanzees had systematically annihilated another, at times with extreme savagery. A few Kahama females of reproductive age survived by assimilating into the attacking group, but all the others died. Similar events have been reported in other chimpanzee groups (Nishida et al. 1985; Ghiglieri 1987, 1988).

These extraordinary observations challenged prevailing views of the chimpanzee social system. The Kasekela and Kahama groups originated from the fission of a single SKIF group in about 1970. Prior to the split, the community had exhibited relative harmony, at least within the volatile chimpanzee definition of that word. The split itself produced two unequal SKIFs, the Kahama emerging much the smaller. Once separated, the groups divided the original territory and adopted the xenophobic stance typical of competing chimpanzees, as if the prior good relations and potential kinship among them had never existed. The Gombe observers had no indication that the relationship would deteriorate further, until the lethal raiding began.

In normal foraging, encounters between chimpanzee groups typically involve loud and raucous displays, sometimes escalating to limited scuffles. However, conflict usually falls short of outright warfare and casualties are rare. At some point in the confrontation, one group retreats, after which both sides return to foraging in their separate ranges. At Gombe, something quite different happened. The larger group systematically invaded the smaller group's territory, attacking the residents they encountered. Observers noted that killings occurred only when the attackers substantially outnumbered the defenders, a situation made more likely by the numerical disparity between the two groups (Manson and Wrangham 1991; King 1976; Goodall 1986). Confrontations between groups of comparable size typically end in stalemate and mutual withdrawal, but in this case, the Kasekela attackers had a numerical advantage. In economic terms, the potential benefit of attack greatly exceeded the possible risk. In such situations, apparently, chimpanzees are capable of murderous aggression. With relentless efficiency, the opportunistic Kasekala group annihilated the Kahama and reclaimed their territory, which they then exploited unchallenged and unencumbered by its prior occupants.

The Gombe scientists continue to study these remarkable events in the hope of finding the causes and precursors of such xenophobic violence. We do not know the degree of kinship between the Kasekala and Kahama, as sexual promiscuity in chimpanzees complicates the assessment of kinship lines. Certainly, the Kasekala behaved as if they recognized no kinship with the Kahama. They obviously gained territory from their action, which improved their resource base. Together, the pressure of resource competition and absence of recognized kinship synergized into a powerful xenophobic force than enabled the Kasekala to perpetrate an atrocity.

Left unchecked, such xenophobic aggression represents a potentially catastrophic path for a species, leading to diminished populations and progressively greater inbreeding. The primary restraining force against such behavior is the evolutionary imperative for genetic diversity. If xenophobic SKIF groups practiced unconditional segregation, permanently excluding all members of all other groups, they would eventually become entirely inbred. The only way for SKIFs to remain genetically diverse is to relax their xenophobic barriers periodically to allow the influx of new reproductive partners. These new members must necessarily emigrate from other SKIFs, perhaps even hostile competitors. Over time, such exchanges of reproductive partners tend to increase the kinship and inlaw connections among neighboring SKIFs, which provides some potential moderation of their natural xenophobia. Hostile competition may even yield eventually to peaceful cooperation and ultimately fusion into a larger SKIF community. Alternatively, as apparently happened at Gombe, xenophobia may regain its strength within the SKIF, leading members to regard all members of the competing SKIFs as outsiders, even those émigrés who still have kinship ties to their former home community.

Some observers of the Gombe events see eerie parallels with human aggression (Wrangham and Peterson 1996). They suggest the chimpanzee studies may be a learning laboratory for understanding human aggression. Others suggest that the events at Gombe were simply the artifacts of the contraction of the traditional chimpanzee range (Power 1991), that they reflect only the dysfunctional society of a species on the brink of extinction. Although chimpanzees and humans have major social differences, we also have similarities. The Gombe studies show us what chimpanzees are capable of. Perhaps such knowledge will help us understand what humans are capable of, and why. The next section deals with the unique patterns of human aggression.

Human Aggression and Control

History amply attests to the human capacity for aggression, violence, and atrocity (van der Dennen 1995; Waller 2002; Hedges 2002). The entire episode at Gombe represents an almost daily occurrence somewhere in the modern human world. From the first recorded history, aggression dominates the human narrative. We know less about prehistoric violence, but there is no reason to believe that the early hominids were immune to violence. Sociologists and psychologists have gone to great lengths to explain human aggression, but consensus has yet to emerge, other than to restate the obvious: it happens and we cannot seem to prevent it. Such deep-rooted problems may indicate equally deep-rooted causes.

Historically, human aggression probably occurred on a small scale, most likely following the same raiding patterns displayed by the Gombe chimpanzees: males engaged in surprise attacks on isolated victims, followed by looting, kidnapping, and retreat (van der Dennen 1995). The primary motivations probably involved expansion of the resource base and the assimilation of reproductive females. In retaliation, the victim's tribe might launch a similar raid, taking females and resources from the previous aggressor. Such reciprocal revenge feuds characterize modern tribal and clan aggression, and may continue for generations (van der Dennen 1995). From time to time, raids might result in injury or death, which in turn might provoke tit-for-tat retaliation. While technically this qualifies as serious aggression, certainly for the victims, it has relatively minimal impact on the stability of the competing SKIF groups.

As long as small hominid groups occupied large ranges, such limited aggression could persist virtually indefinitely. However, as hominid populations grew, pressure on resources necessarily increased. Larger populations needed larger territories. These could be acquired either through migration into virgin land or by the aggressive takeover of territory from neighboring groups. We know that hominids as early as *Homo erectus* migrated long distances (Lewin 1993; Walker and Shipman 1996), an event that might have been motivated by territorial needs. However, migration to new lands was impractical for SKIFs surrounded by competing groups. With no access to virgin territory, but with growing populations to feed and shelter, many hominid SKIFs were forced to compete with one another for ever decreasing territories, much as the Gombe chimps are forced to do. As their competition became acute, the potential for lethal aggression escalated, eventually yielding to uniquely human patterns of warfare, atrocity, and genocide.

To understand how such escalating human aggression might have evolved, consider the following scenario. Two neighboring SKIF groups reach the maximum populations supported by their respective home ranges. To acquire the necessary resources, they must encroach aggressively on each other's territory. Let's suppose that after a period of many years, one group eventually achieves a winning strategic position. The other group capitulates, leaving the winners free to assimilate the new territory.

Victory presents some practical problems, however. On the one hand, the winners cannot consolidate their economic gains until they disperse the losers from the occupied territory. Leaving them on the land to consume resources, and perhaps mount a future counterattack, is not a sustainable outcome. On the other hand, certain segments of the vanquished population, like females with reproductive potential, have great value. Thus, to exploit their victory, the winning group needs to separate males and females in the defeated group, selectively disperse the males to reduce the total population, and assimilate the females into the breeding population.

Chimpanzees would implement this strategy without hesitation. However, the same approach applied to a vanquished human SKIF would entail some major risks. For example, arbitrarily dividing a human SKIF community by gender necessarily entails breaking powerful pair-bonds between males and females, separating parents from their children in whom they have invested heavily, and disrupting recognized kinship bonds, both direct and indirect, that define the inclusive genetic fabric of the SKIF community. This represents a catastrophic social challenge in the human context, and much more likely to engender intense resistance than the comparable task in a chimpanzee society.

In addition, the victors in any territorial aggression face the unpleasant reality that their own home range is a potential target for other invading groups. Expelling conquered males after a successful takeover merely adds to the number of potential future invaders. Plus, the exiled males would have the added motivation of rescuing their wives and daughters and reuniting their families. Thus, the option of exiling unwanted portions of vanquished populations potentially increases future threats.

What is the answer to this dilemma? Obviously, the option of annihilation becomes more tempting. Massacring the conquered males offers the desirable prospect both of reducing the population and of quelling resistance to assimilating kidnapped females. In addition, the victors need not face the same enemy again in future territorial conflicts. By adopting a strategy of systematic annihilation, in effect perpetrating genocide, the victors in any dispute guarantee greater control in the aftermath of their victory. Thus, the logical escalation of perpetual territorial conflicts among human groups passes from limited skirmishing to organized warfare, and then to atrocity and genocide.

Note that the very characteristics that promote cooperation and harmony within SKIF communities represent, when threatened, the flashpoints for catastrophic violence between SKIFs. Our predisposition for pair-bonding represents, on the plus side, a powerful mechanism for nurturing offspring, but on the minus side, a source of violent provocation when threatened. Similarly, the human male's strong parental investment in his offspring will intensify his resistance to threats against them. In simplistic terms, while a man might fight for his land, he will kill for his wife and family. Thus, in human social systems, our negative potential for lethal aggression mirrors, darkly, our positive potential for cooperation and support. One is the price we pay for the other.

If we follow this logic, we must conclude that aggression is a natural part of the human control strategy. In operational terms, aggression is simply action taken to resolve internal dissonance, a natural outcome of a control process. Such actions, like all control-driven behavior, may be desirable or disruptive depending on their social context. For example, a hominid male might hit a leopard with a rock in order to drive it away and scavenge its kill. It is an aggressive, violent act. But if it is successful, it resolves the dissonance associated with his hunger. In this case, we praise the aggressor as clever and innovative in securing a rich supply of food. However, that same individual might then hit another human with a rock in order to kidnap his mate. This is also violet aggression, identical in its physical form to the leopard incident, but in this case directed against a conspecific to resolve sexual dissonance. Here, we are inclined to condemn the aggressor for disrupting social harmony and stability. But in the purely mechanistic context of control, these two scenarios are equivalent: violent actions taken to bring perceptions into line with expectations. The aggressor gains in both scenarios, food in one case and sex in the other, which reinforces the value of the aggressive action as a solution to the basic control problem. In the future, the same individual, confronted with the same circumstances, will be more likely to engage in the same aggressive behavior. Thus, human aggression emerges not as a new form of social behavior, but as just another manifestation of familiar actions that have proven useful to hominids in controlling events over the millennia.

It is the social context, not the action itself, which determines whether violence is appropriate or inappropriate. Focusing on the action without understanding the accompanying perceptions and expectations ignores the fundamental nature of control. Thus, from the aggressor's standpoint, aggression yields control, while from the victim's standpoint, it diminishes control. For some members of the local SKIF group, like the aggressor's kin, the action adds resources to the family. For others, like the victim's family, it diminishes their fortunes. When aggression occurs in a social group, the patterns of dissonance among the various members may change in complex and chaotic ways. Some will consider it a desirable action that brings control; others will see it as an undesirable perception that requires counter-control.

This dual nature of aggression is a conundrum for all social species, but it is particularly acute for humans. Our cognitive structure is finely tuned to titrating dissonance, making innovative use of actions, tools, and forces to bring our perceptions into line with our rich expectations. Our ability to project outcomes forward in time often reveals the short-term benefits of violence, but not always the long-term consequences. Our cognitive minds experience dissonance on many levels, which complicates the calculus of peace and violence. The challenge for our species is to utilize aggression when it is appropriate and to suppress it when it is not. The ultimate arbiter of appropriateness is, of course, natural selection.

Atrocity and Genocide

Viewed in the context of control, aggression has no intrinsic positive or negative value. It is simply action that moves some perception toward some expectation. To deal with human aggression, we need to focus not on the action, but on the human expectation that warfare, atrocity, and genocide are acceptable manifestations of control behavior. While we can readily understand individual violence in defense of home and family, it takes considerable organization and planning to effect genocide (du Preez 1994; Waller 2002; Staub 1989). As Robert Zajonc said of the 1994 Rwandan genocide, "It is quite a leap from inclusive fitness to the slaughter of 800,000 Tutsis in just 100 days" (quoted in O'Toole 1999; Zajonc 2002).

Waller (2002) has outlined the case for multiple cultural and biological factors that promote extreme aggression. One trait in particular emerges from this analysis: the human capacity for *cognitive objectification*. This unique trait gives us the means to transform our enemies from fellow human beings into mere objects of control. It is the first gateway to genocide. Cognitive objectification follows the logic outlined in previous chapters of this book. Put simply, once an external entity is transformed by the observer's brain into a pattern of neural impulses flickering through cortical cell assemblies, it is becomes a virtual object, essentially indistinguishable from all others. Like the glass beads in Hermann Hesse's great novel *Magister Ludi*, the human mind has an almost infinite capacity to reduce the world into collections of sterile symbols that we can arrange at will on a virtual game board. We can put them in patterns, change their characteristics, juxtapose them with other beads, toy with relationships, and so on. We can even destroy them. In the neural milieu, all cognitive objects, human and nonhuman, have the same relative value.

When humans shift from cognitive objectification to the actual objectification of other human beings, individually or in groups, it becomes what sociologists call *dehumanization* (Erikson 1966). This is the second gateway to genocide. Dehumanization is actually something we all do as a matter of course. For example, when driving on the freeway, do we consider the humanity of the driver in the car ahead, or do we simply react to the object in our way? Does our impatience reflect a human conflict with that particular person or simply blind frustration with a thing? In political and military planning, dehumanization happens quite deliberately as people become mere flag pins on a map, literally like glass beads on a game board. As leaders plan, they may move the pins, juxtaposing them with the opposing pieces and analyzing the outcomes. The players know that the symbols represent people, but controlling the board requires treating them as objects. Fortunately, most of us snap out of such dehumanization at the first sign of humanity from our *objects*. A wave from another driver or a salute from a would-be adversary restores their humanity, reconnecting them to some inclusive social system.

Occasionally, however, such reconnection fails and dehumanization persists. In such cases, the sociological term *pseudospeciation* may apply. This is the third gateway to genocide. It refers to the segregation of individuals into groups that cannot reconnect to the larger group. This is conceptually similar to what zoologists do in defining new species, that is, identifying separate groups that have no reproductive interest in one another. Once a group falls *outside* the species, the innate social overlays that operate *within* the species no longer apply. Aggression in this context is no longer social; it is predatory. Persecution, atrocity, and genocide emerge naturally as predatory strategies, and their dynamics reflect an organized hunt rather than a spontaneous outburst of anger or frustration. As long as the aggressor does not allow the victims their humanity, such predation can continue indefinitely. When social control reaches this stage, the aggressors may perpetrate atrocities with no more hesitation than the mundane actions they do in their everyday lives. Get up, eat breakfast, massacre neighbors.

Some argue that the events among the chimpanzees at Gombe represent a prototypical genocide. Goodall remarked that the savagery perpetrated on the victims was comparable to the actions of hunting parties killing monkeys. This argues that humans are not unique in their capacity for pseudospeciation. Clearly, however, the Gombe attacks, while savage, do not approach the scale that human genocide routinely achieves. To paraphrase Zajonc, it is quite a leap from Gombe to the Holocaust.

Some authorities characterize atrocities as failures of human control, but I believe this misrepresents the term. Consider the episode at My Lai during the Vietnam War. At his court martial in 1970, the commanding officer, Captain Earnest Medina, faced charges of not having exercised proper control over his troops. In fact, the testimony showed My Lai to be a study in control. The troops moved events forward with a deliberate efficiency that matched the expectation they had allegedly received from their officers. A group of otherwise normal American men acquired a shared expectation, and then acted cooperatively to bring their individual perceptions into line. Never mind that the process excluded the humanity of the victims, and ignored the objective realities of the military situation. Once that horrific expectation settled in their minds, the reduction of dissonance motivated their aggressive behavior, exactly as one would predict from control theory.

Homo bellicosus and the Human Strategy

Lethal aggression among humans may not be unique in the primate world, but in its quantitative excess it is a distinctive, if unenviable, human trait.

To the degree that it connects to our drive for control, then it must be a natural by-product of our success as a species. While we would like to believe that murder is maladaptive, natural selection does not deal in belief, only reproductive success. If *Homo bellicosus* endures, while less aggressive human variants do not, then its violent nature shows us our future. Alternatively, if *Homo bellicosus* is an aberration, a temporary variant ultimately maladapted to long-term change, then the horrors of war and genocide may eventually disappear. The propensity of *Homo bellicosus* to destroy entire gene pools, as in the Nazi vision of Aryan dominance, argues that natural selection must ultimately work against them. But how much warfare and genocide must we endure before we escape such temptations? If genocide gives us the promise of control, then what, if anything, will ever stop us?

The emergence of *Homo bellicosus* illustrates the potential consequences of a control strategy unchecked by moderating influences. Human violence is both a control solution and a control problem. It remains in our behavioral repertoire because it gives us the control we require in certain situations. It has not yet consumed us, because it is itself subject to control. In the next two chapters, I suggest two restraining mechanisms, altruism and cultural signaling, which natural selection has installed as biological governors on human aggression. Their restraining influences, I believe, protect us against the runaway aggression of *Homo bellicosus*. They offer the hope that human violence, atrocity, and genocide may eventually subside into distant memory. Without some kind of restraint, *Homo bellicosus* would surely have killed us all long ago.

CHAPTER 7

Homo beneficus, the Altruist: The Kindness of Strangers

Bang the Drum Slowly

In his book *Bang the Drum Slowly*, Mark Harris (1956) explores the interplay between selfishness and altruism against the backdrop of professional baseball. The main character, Henry Wiggin, a talented pitcher, befriends a catcher named Bruce Pearson. Wiggin learns that Pearson has Hodgkin's disease, a fatal lymphatic cancer. Pearson has only marginal ability, which his debilitating disease certainly cannot help. The team has an opportunity to win the pennant, which would add substantially to the players' paychecks and prestige, but Pearson's poor play could possibly cost them the championship. Against his own personal interests, Wiggin stands by his friend, defending Pearson against the attacks of teammates and management. In his unique way, Wiggin makes a case for his altruism:

> "Probably everybody be nice to you if they knew you were dying," [Pearson] said.
> "Everybody knows everybody is dying," [Wiggin] said, "That is why people are nice. You all die soon enough, so why not be nice to each other?" [p.140]

Then later in a conversation with Pearson's gold digger girlfriend:

> "Then why not live it up a little?" [Katie] said. "Why worry so much about Pearson's old man?"
> "I do not know," I said, and that was true, for I did not. Do not ask me why you do not live it up all the time when dying is just around the corner, but you don't. You would think you would, but you don't. "I do not know why," I said. [p.217]

In the end, given his friend's support and the opportunity to play, Pearson does better than expected and helps the team advance to the World Series. But the effort weakens him and he succumbs to his illness. Wiggin is the only member of the team to attend the funeral.

Harris's fictional ballplayer offers us a simple, yet profound, view of altruism: that we just do it and no one knows why. Logic seems to favor a selfish strategy: that with dying just around the corner, it pays to be selfish. Yet humans still have the propensity to behave altruistically, a trait that social scientists have struggled to explain. With our cognitive ability to project events into the future, to see the eventual consequences of our actions, why do we ever choose to help each other? Knowing our fate, and that of everyone around us, why does anyone ever do the right thing?

In Chapter 6, I asked a similar question, but in a darker context. If killing people gives us the control we crave, then why stop? Why negotiate or cooperate with others when homicide might solve the problem with one swift stroke? Lethal aggression, the simple logic of *Homo bellicosus*, is the ultimate expression of selfishness. In a world of limits and looming mortality, why should we ever pull back from such behavior?

In Henry Wiggin's view, we just do. In social scientific terms, we have an innate predisposition toward altruism. This term refers generally to actions that enhance another's fitness, at some cost to us, and without necessarily, directly, or immediately enhancing our own fitness. Altruism contrasts with selfishness and aggression, actions that promote purely personal interests.

Social scientists have hotly debated the existence of true altruism for many years. The sociobiology school tends to view it as a phantom, that the appearance of *unconditional* self-sacrifice always involves some selfish gain. They argue that if true altruism arose from certain variant genes that invaded and expanded in the human gene pool, then according to evolutionary theory, such genes must have had adaptive value to the ancestors that first expressed them, in effect conveying a reproductive benefit to the first so-called altruists. In the minds of some sociobiologists, such fitness gains necessarily taint the notion of altruism as pure self-sacrifice.

At the other extreme, the cultural school asserts that true altruism can exist as a learned tradition shaped by social conditioning in morality and righteousness and modeled by moral leaders. This view rescues altruism from the rigors of the selfish gene, but leaves it as a purely cognitive construct, an acquired behavior maintained by perpetual training. While this may reflect the prevailing human experience in most social systems, proponents of this view are hard-pressed to explain how such conditioning persists against the potential invasion of learned selfishness. How is it that generations of humans retain altruistic habits when they could just as easily abandon them and substitute selfish habits with potentially greater personal benefit?

Thus, the problem with altruism is that neither model, genetic nor cultural, can produce a compelling explanation of how self-sacrifice can endure against self-interest. This conundrum leaves most behavioral biologists searching for the reality of altruism somewhere between the two extremes. While unconditional self-sacrifice may be rare in real societies, so is unmitigated selfishness (Gintis et al. 2003). Somewhere between the cold mathematics of selfish genes and the warmer, but inaccessible, realm of moral intent lies the truth of human altruism.

A Perfect World

We can illustrate the paradox of altruism with a thought experiment. Imagine a species living in a perfect world where all its needs are met without cost or competition. Food is plentiful, predators are absent, reproductive partners are ubiquitous. Suppose this Garden of Eden extends from horizon to horizon and endures unchanged through time. After thousands of generations in such a world, would the indigenous species be altruistic? The answer, I believe, is no. If individuals have everything they need, without cost or restriction, then self-sacrifice is unnecessary and irrelevant. Indeed, social cooperation in general is unnecessary in such a world. Absent the selective pressures of starvation, predation, and reproductive failure, there is no apparent adaptive value in altruistic behavior. No sacrifice is necessary if no problem exists. It follows, then, that in a perfect world, true altruism does not exist.

If the inhabitants of a perfect world are not altruistic, then are they selfish? Here again, the answer is no. In a perfect world, there is likewise no basis for selfishness. For example, the act of stealing food is meaningless if food can be replaced without cost. If no predators threaten, then selfishly shirking one's defensive duties has no meaning or consequence. Thus, in our perfect world, the notion of selfishness also disappears. To the inhabitants of such a world, neither selfishness nor altruism has any meaning.

Consider now the real world of uncertain food supplies, of lethal predators, parasites, and pathogens, and limited reproductive opportunities. In such a world, obviously much closer to the reality of the human evolutionary environment, what are the relative prospects for altruism and selfishness?

Selfishness clearly has a payoff in a world of limitations. If food is scarce, for example, then stealing it from others represents a rich new source. Thieves eat while victims starve. Similarly, when facing predators, the selfish act of abandoning a weaker individual may constitute a successful defense. The predators attack the hapless victim, leaving the others to escape. In the struggle for reproductive rights, a selfish bully who secures the largest harem has a clear reproductive advantage. Thus, in a world of limitations, selfishness emerges as a natural survival strategy.

The key question is whether the world of limitations also promotes the emergence of altruism. The answer, I believe, is yes, but it happens in an indirect way. According to sociobiological theory, true altruism has virtually no chance of invading a selfish population or persevering against the invasion of selfishness (Dawkins 1976; Hamilton 1964; Alexander 1974). The sacrifices made by altruists leave them open to exploitation by selfish individuals, eventually rendering them deprived, increasingly unfit, and at risk of extinction. In head-to-head competition, selfishness inevitably wins over true altruism, according to most sociobiological models. Thus, we should conclude that the real world cannot produce and sustain true altruism in its social systems. However, the sociobiologists may be asking the wrong question. It is not whether selfishness defeats altruism, but rather what happens to selfish human societies that fail to develop altruism.

The story of *Homo bellicosus* (Chapter 6) gives us a clue. The critical question raised there was how a species so utterly focused on controlling events could avoid lapsing into runaway aggression and genocide. If aggression is a logical outcome of selfishness, and both reflect the compelling human instinct to control events, then the fate of selfish societies is precisely the withering aggression depicted in Chapter 6. Unchecked aggression has profound evolutionary consequences: it reduces reproducing populations and it depletes genetic diversity, both significant risk factors for extinction. Thus, a human society that veers into unmitigated selfishness incurs a greater likelihood of lapsing into unchecked violence, which leads inevitably to extinction. Thus, while most sociobiological models might predict the victory of selfishness over altruism, they should also demonstrate that selfishness, unrestrained, must also increase the likelihood of extinction. While the selfish species might endure long after the altruistic ones are gone, in the end, they cannot avoid their own self-destruction, and they too will disappear. Societies based purely on the pursuit of selfish interests are evolutionarily unstable.

Clearly, the human species has not succumbed to such extinction, despite our demonstrated propensity for selfishness and aggression. This suggests that there may be mitigating factors pulling us back from the abyss. Altruism may be one such. To be adaptive, it does not need to eliminate the negative traits of selfishness and aggression; it only needs to check their self-destructive spiral. Clearly, if humans had a lower propensity for lethal aggression, we would have less need for altruism, as in the perfect world scenario above. But in our imperfect world, where aggression is a primary tool of control, altruism represents a natural and necessary counterbalance, a moderating trait that prevents us from going too far. Put simply, altruism does not make us good; it simply prevents us from becoming evil.

In this view, altruism and aggression are co-evolutionary social traits. In a world of limitations, the members of one's own species can be either cooperative partners or threatening competitors. Competition pushes societies apart, reflecting the logical consequences of our species' natural tendency toward selfish control. Altruism moderates this repellent effect, biasing social systems toward cooperation. In every society, there is a dynamic tension between cooperation and competition, between altruism and aggression. In some situations, altruism dominates to produce highly cooperative interactions. In others, altruism fails, leading to intense competition, sometimes with horrific consequences.

In Chapter 6, I defined various categories of social interaction in terms of their aggression potential. By the co-evolutionary logic, each of these elements must also have an altruistic dimension. Figure 7.1 summarizes this concept, using the same basic schematic as our aggression model in Chapter 6. Together, these components define a composite tendency toward altruism in our species. The following sections describe each one. As we shall see, a measure of altruism can promote harmony and cooperation within human social systems, but it is often insufficient to prevent aggressive outbursts, particularly between xenophobic communities.

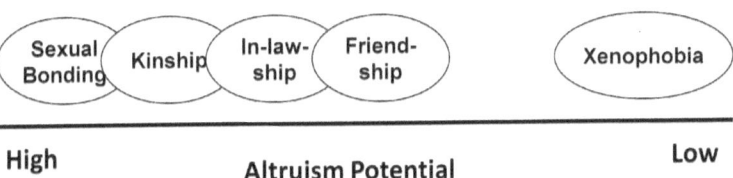

High   **Altruism Potential**   Low

Figure 7.1. Human social components mapped on the dimension of Altruism Potential. This map is identical to that of Figure 6.1, except the High-Low axis is reversed. Sexual interactions represent the system most likely to demonstrate altruistic behavior. Kinship, inlaw, and friendship systems display progressively less potential for altruism under normal social circumstances. Xenophobia again defines the extreme end of the spectrum, reflecting the virtual absence of altruism between competing social groups. The similarity of the altruism and aggression maps may reflect their counterbalancing roles in social structures, and perhaps their co-evolutionary origins.

Sexual/Reproductive Altruism

Sexual reproduction is the prototypical social system, upon which all others must build, and for which they must provide support. Sexual behavior also incorporates a prototypical form of altruism. Consider that it requires two individuals to cooperate in a complicated behavioral ritual that renders both vulnerable to attack by the other, that potentially exposes both to attack by rivals, predators, and pathogens, and that apparently does not produce any immediate benefit. In primates, the reproductive benefit of sex may not accrue for many months or years, yet the risks occur immediately. Contrast this with nonsexual reproductive modes, such as parthenogenesis, that do not expose the individual to social risks. Similarly, in nonreproductive behaviors, such as feeding or sheltering, selfish action brings benefits with minimal delay. Compared to these other behavioral repertoires, sexual reproduction appears to carry significant costs without immediate compensation.

Species vary widely in their displays of sexual altruism. In some, the reproductive commitment lasts only through egg deposition and fertilization, after which both male and female disappear, leaving their progeny to their fate. Both male and female accept the effort and risk of reproduction, but not much more. Other species, like primates, display considerably greater reproductive altruism, engaging in extended behavioral rituals, often differentiated by gender. Males may invest considerable time and energy in prenuptial activity, in some cases taking major physical risks in order to gain access to females. Once fertilization occurs, however, they disappear quickly, leaving the remaining burdens of gestation, birth, and nurturing entirely to the female.

Humans, in contrast, display stronger reproductive altruism and more balanced gender roles. Human reproduction differs from that of the other great apes in the formation of enduring pair-bonds between the reproductive partners. For the human male, this demands a relatively greater predisposition toward reproductive altruism. Arguably, pair-bonding compromises the human male's overall genetic fitness. If he simply abandoned the pregnant female, and copulated with as many other females as possible, he could theoretically transmit his genes to more offspring in a shorter time. Chimpanzee males pursue this strategy vigorously, while human males are apparently less inclined. This suggests a fundamental evolutionary shift. Its root cause is probably the unique burden of human parenting associated with increased infant helplessness, as discussed above in Chapter 5. Without dual parenting, human infants face significantly greater mortality rates (van Schaik and Dunbar 1990; van Schaik and Kappeler 1997). The death of an infant clearly punishes both parents' reproductive fitness. Thus, from a reproductive standpoint, selfish males, who abandon their pregnant partners to pursue new females, may not gain significantly against pair-bonded males who forego such philandering. If human infants were more robust, the pair-bonding strategy might never have developed among the emerging hominids, and our divergence from the chimpanzees would be less salient.

Pair-bonding implies a greater predisposition toward altruism in the human species, but it is not entirely unselfish. While both partners sacrifice other reproductive opportunities in this arrangement, the compensation comes in the increased survival and health of their progeny with the attendant improvement in their future reproductive success. Thus, even though pair-bonding demands sacrifices that do not appear to produce any short-term return, it may yield a higher long-term probability of reproductive success.

Human pair-bonds are highly cooperative and relatively noncompetitive, despite the fact that the participants have little, if any, genetic kinship. Each partner retains the capacity for aggression, but the powerful altruism that underlies the relationship tends to immunize the partners against lethal violence. In some situations, of course, this protective mechanism can break down and aggression can erupt within the pair-bond. Obviously, however, this maladaptive outcome does not promote reproduction. Practiced widely within the species, it would increase the risk of extinction. More often, reproductive altruism produces relatively harmonious and enduring pair-bonds that represent the cornerstone of human society.

Kinship Altruism

Reproductive success brings the birth of offspring and the emergence of *kinship altruism*. This form of altruism reflects the bonds between family members who share a substantial portion of their genes. It differs from reproductive altruism, which is a cooperative relationship between unrelated individuals.

Kinship altruism is a behavioral reflection of *inclusive fitness*, discussed in Chapter 6. The theory of kin selection (Haldane 1955; Hamilton 1964) says that natural selection will favor behavioral repertoires that enhance the fitness of relatives who share one's own genes. For example, defensive behavior intended to protect close relatives has the same theoretical adaptive value as self-defense. Both increase or maintain the frequency of shared genes in the population. Similarly, behavior intended to promote the reproduction of closely related kin should promote one's own genetic fitness to some degree. Carried to the extreme, the willingness to sacrifice one's own life to assure the survival and reproduction of multiple kin could produce a net gain in the genetic fitness of the species, as viewed strictly from the evolutionary perspective. Although theorists apply the term altruism to this pattern, it eventually serves the selfish interests of genetic perpetuation.

Kinship altruism takes its prototypical form in parental investment in progeny. Both parents have kinship with their progeny, so both benefit genetically from their survival and eventual reproduction. Kinship altruism also encompasses supportive behavior among close family members such as siblings, cousins, aunts, uncles, grandparents, and so on. In an extended family, each member carries genes in common with the other members. The degree of kinship depends on the number of shared genes, so we have fewer in common with our cousins than with our siblings, but more in common with our cousins than with unrelated individuals. The theory of kin selection says that altruistic behavior within the kinship group has adaptive value in direct proportion to the degree of relatedness.

Kinship altruism remains an enduring force even as offspring mature and establish their own reproductive pair-bonds. As they produce their own offspring, siblings and parents may remain in close proximity, assisting in provisioning, defense, and childcare. Thus, starting from the original pair-bond, kinship groups grow ever larger, including aunts, uncles, and cousins. The original pair becomes patriarch and matriarch of an emerging clan. Kinship altruism promotes the continued pooling of resources and support within this growing group.

The altruism that underlies kinship groups restrains its members from engaging in unchecked aggression against one another. Family members may squabble over status or possessions, but they rarely resort to lethal violence. Kinship altruism moderates such aggression in direct proportion to the degree of kinship. The prohibition is much stronger within the nuclear family than in distant relationships. As kinship diminishes, the power of altruistic restraint decreases and the propensity for competition, including lethal aggression, increases.

Inlaw Altruism

If kinship groups remained isolated, breeding incestuously over the generations, they might eventually achieve a level of kinship altruism more powerful than any observed in nature. However, in evolving species, this cannot happen. The evolutionary requirement for genetic diversity requires the continual infusion of unrelated individuals into the growing clan. To accomplish this, human social systems must have mechanisms for attracting and assimilating unrelated individuals into a kinship group. This in turn entails establishing cooperative relationships between the two kinship groups, a process I call *inlaw altruism*.

Inlaw altruism represents a social bridge between unrelated and otherwise competing kinship groups. It differs from kinship altruism in that émigrés share relatively few genes, if any, with their new groups. It also differs from reproductive altruism, which applies only within the pair-bond framework. Inlaw altruism defines the relationship between an émigré who pair-bonds with a member of a kinship group and the other members of that group. This relationship becomes manifestly supportive once the émigré becomes the parent of offspring that have direct kinship to the group. Under the principle of inclusive fitness, it is in the group's genetic interest to support the offspring of related group members. Logically, it is also in their genetic interest to support the parents of such offspring even though one of them may be genetically unrelated to the kinship group. Failure to support the exogamous parent may disadvantage the progeny, which works against the group's inclusive fitness. Thus, the mechanisms of kinship altruism must somehow generalize to include unrelated individuals. Inlaw altruism underlies the cooperative interactions of inlaw social systems, as outlined in Chapter 6.

Following this logic, inlaw altruism should also extend further to include *both* kinship groups that share the pair-bonded partners. Under inclusive fitness, both kinship groups benefit directly by supporting the pair-bond and its offspring, so it follows that both benefit indirectly by supporting each other. Once the pair-bond produces offspring, mutual inlaw altruism between the kinship groups supports their shared gene pool. Thus, the extension of indirect kinship altruism to include interactions between the originating kinship groups has some selective advantage under the principle of inclusive fitness.

The recognition and acceptance of inlaw relationships may extend to the merging of kinship groups into larger social cooperatives or clans. Although the generalization of altruistic behavior beyond the boundaries of pure kinship still rests indirectly on the genetic foundations of inclusive fitness, the extension of cooperative support to unrelated individuals represents a major shift in human social evolution. Without it, exogamous pairings might face abandonment by their originating kinship groups. This would leave them less fit for successfully rearing children, a problem made more acute in the human species by the greater helplessness of human infants. Thus, the evolution of inlaw altruism helps preserve the links between kinship groups, providing a more supportive environment for children.

Although inlaw altruism should promote cooperation between kinship groups, it is clearly less powerful that kinship altruism itself. While it may moderate competition between kinship groups, it may not have the power to eliminate it. Depending on circumstances and experiences, families may lapse into aggressive competition, sometimes with tragic consequences. Recall, for example, the classic unhappy ending of Shakespeare's *Romeo and Juliet*.

Inlaw altruism does not prohibit aggression between groups. It provides only a restraint, weak and indirect, that retards the descent into violence and its maladaptive consequences for the offspring shared by the families. The unique parenting needs of the human species benefit from this form of altruism as hedge against aggression, but it does not guarantee a harmonious outcome in all relationships.

Friendship Altruism and Reciprocity

Extending the concept of indirect kinship still further, altruism between unrelated individuals should also exist whenever their relationships build on shared family interests. An outsider who protects a friend's family promotes that friend's genetic fitness, just as true kin would. It follows that altruism extended to such individuals should promote genetic fitness in the same way as kinship altruism. I use the term *friendship altruism* to describe this behavior.

The altruism displayed in friendship relationships is weaker than that among inlaws since there is no shared genetic component. Although friendship linkages may be quite strong and enduring, and may eventually become inlaw relationships, unequivocal displays of altruism among friendship groups are less likely than among extended families.

Social theorists suggest that unrelated social partners adopt the operating principle of *reciprocal altruism* to stabilize their partnership (Trivers 1971; Axelrod 1984; Axelrod and Hamilton 1981). This theory holds that unrelated individuals may act to benefit each other, at some cost to themselves, if the reciprocal benefits and costs exactly balance over time. This principle extends to all aspects of social behavior, such as resource allocation, defense, reproductive exchanges, rearing children, and so on. Its emergence in complex social systems represents a way for unrelated individuals to behave cooperatively in a stable social system.

Theorists argue whether reciprocity constitutes true altruism or simply carefully balanced selfishness. There is no doubt that reciprocity represents a powerful mechanism for promoting cooperation in social systems. Individuals who violate the perceived reciprocal balance risk the loss of resources and support, leaving them socially isolated. As a result, they may experience increased predation, resource shortfalls, xenophobic attacks, higher infant mortality, and so on (Alexander 1974; Slurink 1993).

Unlike kinship altruism, reciprocal altruism in friendship circles requires a mutually agreed system of value among the participants. Reciprocity demands that we attach a fair exchange value for our respective sacrifices. For example, I give you a pound of meat; you give me a basket of fruit. Your sacrifice in finding and gathering extra fruit must balance my sacrifice in procuring an extra supply of meat. This requires a basis for computing the relative values of fruit and meat, as well as the relative effort of producing them. Clearly, this implies a degree of economic sophistication, without which reciprocity can go no further than the most basic exchanges.

Humans are unique in their cognitive capacity for computing intrinsic value, as we explored in Chapter 5. The concept has its origins, at least in part, in the enhanced ability of the early bipedal hominids to transport goods from one location to another. By liberating objects from their original locations, improved transport unveiled the concept of intrinsic value, and thereby enabled the economic activities of accumulation and distribution based on value. Given the very early origins of bipedalism, the hominid predisposition toward reciprocity may have a long history.

The concept of economic value also parallels the emergence of cognition. Goodall (1986) noted that chimpanzees demonstrated a limited form of reciprocal altruism where individuals expected repayment for good deeds and cheaters received punishment. However, the chimpanzee economy clearly functions at relatively modest levels. They display little understanding of abstract value. In the early hominids, however, the potent combination of object mobility and cognitive abstraction enabled complex economic computations and complex systems of reciprocity (Seabright 2004).

We see perhaps the most powerful evidence of reciprocity in the division of labor. Human societies exhibit a wider diversity of individual social responsibilities not generally seen in other species. Chimpanzee social systems, for example, show a limited division of labor. Females clearly take a much greater role in infant care, while males accept a heavier load in defensive patrolling and meat procurement. However, a typical day in a chimpanzee group sees most members doing similar activities, such as foraging for food, napping, grooming, and so on. Individuals with tool-using skills may employ these in their own foraging, but they do not generally share their yield with the larger group.

In contrast, human societies tend to divide into specialty groups, each focused on specific tasks whose output adds to the aggregate group economy. Labor specialization exploits the natural variations in human abilities, leading to more productive group economies. In any group, individuals will vary in their ability to do various tasks. Some may show greater courage and strength as hunters, others will have skill as toolmakers or basket weavers, still others as fishers or trackers or foragers. The principle of reciprocity enables individuals to spend a disproportionate amount of time doing what they do best, and still obtain the goods and services they need through reciprocal trade. Thus, toolmakers supply the hunters with blades, and in exchange, the hunters provide the toolmakers with meat. The basket weavers provide the fishers with better seines and the fishers repay the weavers with a share of the catch. The toolmakers no longer need to hunt or fish, while the hunters and fishers no longer need to make tools. The division of labor promotes greater economic productivity in which all benefit.

Over the course of hominid evolution, the value computations underlying the system of reciprocity grew increasingly more complex. In particular, certain exchanges demanded a *time* computation, that is, a factor related to the payback schedule. For example, hunters could deliver meat only periodically, while foragers and toolmakers could deliver plant products and tools almost continuously. If the gatherers and toolmakers wanted meat, they had to feed and provision the hunters in advance, then wait for a future payback. This meant that their initial sacrifice went unrewarded for some time. They also had to accept that the risk that the hunters might fail, as hunting is an uncertain enterprise. Clearly, then, the emerging system of reciprocity must somehow have accommodated delays and uncertainty in the payback. Psychologists are familiar with this concept as the *deferral of reward*. It indicates a level of reciprocal altruism that only the human species has achieved.

The human predisposition for friendship altruism provides the glue that holds the system of economic reciprocity together. The genetic capacity to treat unrelated individuals as virtual kin underlies the broadening of social cooperation into a full-fledged value economy. Without friendship altruism, that is, trust between partners, the willingness to defer rewards breaks down quickly under the pressure of selfishness. Arguably, only the human species has the cognitive capacity to sustain a complex economic system based on trust in future deliverables. While other species show evidence of friendship, and some display rudimentary reciprocal altruism, only the human species has realized the adaptive benefit of a robust economy built on these simple foundations.

Xenophobia and Group Selection

The various forms of altruism outlined above operate primarily within SKIF communities. Unfortunately, they disappear at the boundaries between communities. Except for occasional exogamous exchanges, the failure of altruism to cross territorial borders allows hostility and violence to dominate the social agenda.

Intractable xenophobia has given rise to a unique theory of altruism based on the principle of *group selection* (Wynne-Edwards 1962; D. Wilson 1975; D. Wilson and Sober 1994; Sober and D. Wilson 1998; Boehm 1999). This controversial theory asserts that natural selection sometimes operates at the *group* level rather than at the individual level, which traditional Darwinian

Theory emphasizes. Group selection theory says that certain genetic traits may persist in a population according to their contribution to the overall health and survival of the group, rather than to any individual reproductive benefit. For example, social groups that contain a high proportion of altruists should always be stronger and healthier than comparable groups composed entirely of selfish egotists. Over time, the group selection theory argues, the reproductive fitness of the overall population improves and the proportion of true altruists in the population increases.

For certain social concepts, like altruism, group selection theory has obvious appeal, as illustrated by the following scenario. Suppose a SKIF community happens to produce certain genetic variants with a strong predisposition for true altruism. These individuals behave selflessly, consistently doing more that their share in all activities. They show greater bravery in repelling predators and expend more energy hunting. They forage for longer periods at greater distances, producing larger quantities of food or materials, which they share generously with the group. Other members of the SKIF community clearly gain from this behavior, enjoying benefits beyond those generated by their own efforts. The altruists, on the other hand, may suffer. They may sustain wounds during a hunt or collapse from fatigue after extended foraging. According to group selection theory, the net improvement of the overall group fitness outweighs the risks to the individual altruists. The healthier group continues to produce altruists until the trait becomes universal in that SKIF community.

The sociobiology school disagrees (Williams 1966; E. O. Wilson 1975; Knauft 1989), arguing that the evolution of altruism occurs only through individual selection, that the fitness of the group has evolutionary meaning only as it reflects the genetic fitness of the individuals who comprise it. For example, in the above scenario, nothing actually promotes the expansion of altruistic genes in the population. The altruist's sacrifices (wounds, fatigue, and so on) continually reduce their own reproductive fitness. If they fail to reproduce, their altruistic genes disappear from the population. Thus, the altruist's acceptance of higher risk keeps them forever in the reproductive minority and always on the verge of extinction.

Moreover, true altruists are vulnerable to parasitic freeloaders who exploit them. Selfish freeloaders may siphon off the surplus meat produced by the altruist hunter. Only the freeloaders enjoy a fitness increase, allowing them to reproduce and pass on their traits for selfish parasitism. Meanwhile, the victimized altruists edge closer to extinction. In population models, selfish freeloaders tend to overwhelm pure altruists, unless severe penalties accrue to selfishness (Gintis et al. 2003). Thus, the sociobiologists argue, group selection for altruism never gets a foothold. It is always prone to extinction under the pressure of selfishness (Williams 1966).

In response, group selection theorists have identified certain situations that favor their hypothesis. Specifically, they assert that group selection is favored whenever competition between social groups (1) pits one group directly against another and (2) raises the penalty for defeat to the extinction of the entire group. This is precisely the *annihilation scenario* outlined in Chapter 6. When groups engage in genocidal competition, the emergence of true altruists in one population or the other may produce a significant competitive advantage. A cadre of selfless warriors fighting without concern for themselves (e.g., the 300 Spartans) could theoretically defeat a much larger enemy of selfish individualists. Their victory, coupled with the annihilation of the losers, would perpetuate the population that produced true altruists, and eliminate the one that did not. Even though many altruistic individuals might die in the conflict, the base population avoids annihilation and continues to produce replacements for the lost altruists. As long as the genocidal competition continues, the steady production of altruists by the surviving population continues to convey an advantage. Any lapse in the replacement of lost altruists eventually reduces the group's competitive advantage, leaving it increasingly vulnerable. Thus, we have the paradoxical result that *true* altruism may arise only in a world of perpetual genocidal aggression.

While group selection theorists might support this seeming paradox, other social scientists demur. First, there is some question whether such harsh genocidal competitions are actually representative of human history. It is not clear that human groups in general have practiced prolonged genocidal competitions. In instances where they have, it not clear that altruism always emerged as the best adaptation for winning. Second, the group selection scenario implies that such altruism can endure only under conditions of *perpetual* between-group competition. In other words, when peace comes and the threat of annihilation wanes, the true altruists have no cause for which to sacrifice. Thus, again paradoxically, only by guaranteeing perpetual conflict can true altruism endure. Third, as peace settles in, the pressure to revert to selfish strategies should increase, leaving the true altruists vulnerable to social parasitism by selfish freeloaders. Fourth, as with previous criticisms of altruistic development, this form of altruism still has no guaranteed path to dominance in the population. Proven altruists lost in battle have no direct way to reproduce themselves. Their replacements must come from reproductive pairs who have either survived or avoided the conflict. But as survivors or noncombatants, they are technically unproven as true altruists. Thus, there must be some mechanism resident in the gene pool that spontaneously produces replacement altruists from a breeding population of unproven altruists.

Lacking this constant infusion, the sacrifices made by the true altruists still leave them genetically disadvantaged and in danger of extinction.

Group selection as a basis for altruism remains an active area of debate. The human propensity for lethal violence is a favored mechanism for serving our control interests, which may create the harsh conditions necessary for group selection (Alexander and Tinkle 1968; Alexander 1974; D. Wilson 1975; Boehm 1999). Perhaps in our history as a species, our controlling disposition produced prolonged periods of violence that created a unique form of *warrior altruism* that promoted *both* unconditional self-sacrifice and genocidal violence, the former directed towards one's group and the latter towards other groups. This bipolar trait might serve as a useful model for the emergence of the conjoined phenomena of nationalism and militarism and among human groups. While we may condemn the xenophobia and violence inherent in such social competitions, they may be, paradoxically, the only source of true altruists in our species.

Cultural Altruism

While sociobiology focuses on possible genetic foundations of altruism, other social theorists argue that true human altruism exists primarily as a cultural phenomenon. Parents teach their children to behave altruistically, those children in turn teach their children, and so on. Altruism thus endures as a human cultural tradition dating back to the first hominids to originate the concept. Over time, this altruistic tradition becomes part of the group's history, ideology, and theology. It endures because each generation learns it from the previous generation, and teaches it to the next.

While the purely cultural view of altruism has plausibility, it also has flaws, notably its vulnerability to the invasion of cultural selfishness and aggression. Cultural traditions are just as susceptible to selfish parasitism as genetic mechanisms. Selfishness can propagate as a learned cultural tradition just as effectively as altruism. Individuals may practice it covertly, deceptively parasitizing altruists and reaping the benefit. Unless there are cultural mechanisms for overcoming selfishness, altruism may not endure any better as a cultural tradition than it does as a genetic trait. The threat of selfishness always looms, and all things being equal, selfishness always wins. Purely cultural models therefore do not resolve the problem of how altruism emerges in a human society any better than purely genetic ones.

The Control Interpretation of Altruism

In previous chapters, I have suggested that the propensity for control differentiates the human species from all others, and that the uniqueness of human behavior derives primarily from that overarching drive. The phenomenon of altruism is no exception. Control, by our previous definition, describes a specific relationship among three fundamental elements: perception, action, and expectation. We infer control whenever action alters perception to match a preestablished expectation. Leave any element out, or change the pattern of their interactions, and the process does not appear to us as control, regardless of context or species. Conversely, include all the required elements in the specified pattern and it appears to us as control, again regardless of context or species. If the species in question is human, specifically *Homo beneficus* in this case, then we must account for its emerging propensity for altruism as a manifestation of underlying control processes. How then does altruism fit within the mechanics of biological control as we have defined it?

There are many answers this question, but perhaps the simplest explanation of altruism in mechanistic control terms lies in the selective inhibition of certain options on the action side of the basic control process. This is illustrated in Figures 7.2 and 7.3. In Figure 7.2, we see the same basic control schematic depicted in previous chapters. As before, this figure shows the functional interrelationships among perception, action, and expectation. Each of these elements has complex neurological correlates in real brains, although we cannot yet point to them in the neuroanatomy. The specific elements displayed include 1) the *expectation* of "No Intruders" in the individual's perceptual range; 2) the *perception* of a hominid "Intruder" trespassing on this range, clearly a mismatch with the preestablished expectation; and 3) a set of potential actions (*Yell, Hit, Kill*) which will transform the perceptual situation in certain ways. To the extent that such actions may resolve the mismatch between perception and expectation, the individual has the means to achieve the desired control of the situation. In Chapter 6, I suggested that the repertoire of actions available to the hominids, *Homo bellicosis* in particular, included various forms of violent aggression, including lethal force. That *Kill* option is included in Figure 7.2. Depending on circumstances, the option of killing the intruder may be the simplest and surest way to resolve the perceptual mismatch. From a control perspective, this murderous act produces the desired perceptual state and the individual system achieves stability.

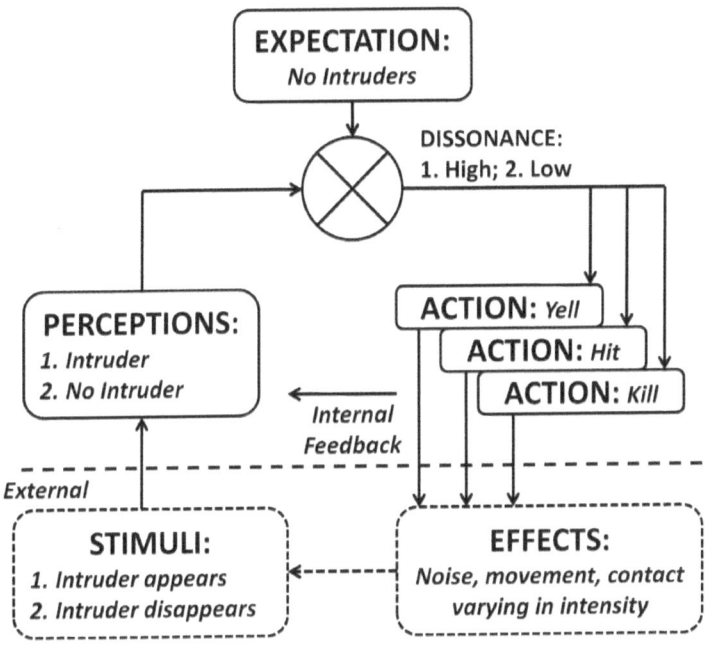

Figure 7.2. Hypothetical control mechanisms underlying hominid aggression and altruism. This figure depicts the case where there is no altruism between the conflicting parties (compare to Figure 7.3). An individual experiences dissonance when a perception of an "Intruder" creates a mismatch with the expectation of "No Intruders." The possible actions range in intensity from nonlethal to lethal. Possible outcomes include flight, injury, or death of the intruder. Action continues until perception matches the "No Intruders" state. In this scenario, the murder of the intruder is a permissible outcome.

The problem with lethal violence, as we saw in Chapter 6, is that it can lead to catastrophically dysfunctional social systems that leave the species more vulnerable to extinction over the long term. The necessary restraint, I have suggested in the present chapter, comes from the subtle, but steady, tug of altruism on the control systems of *Homo bellicosis*. Figure 7.3 shows one possible version of that restraining force. Here, we arbitrarily define an inhibitory function, which I call an *altruistic overlay*, which selectively tamps down the more violent options for action, while leaving the less destructive options. This overlay responds to certain perceivable attributes, such as recognition of the intruder as a member of the local SKIF community. When activated, this overlay alters the dissonance computation, and thereby selectively suppresses option of lethal violence. Various nonlethal actions remain available, including bluff and bluster, verbal abuse, and limited physical punishment, but the altruistic overlay prevents these options from escalating into lethal intensity. The behavioral repertoire thus modified is consistent with the pattern of altruism. Even though the supposed *altruist* still does not tolerate intruders, she selectively withholds extreme behavioral options. This may not seem overly altruistic, but compared to the full range of lethal alternatives available to *Homo bellicosus*, it represents a considerable step forward. Of course, the altruistic overlay is active only in selected circumstances. If its inhibitory restraints ever weaken, the option to engage in lethal violence will instantly return. Its targets could even include members of the local SKIF.

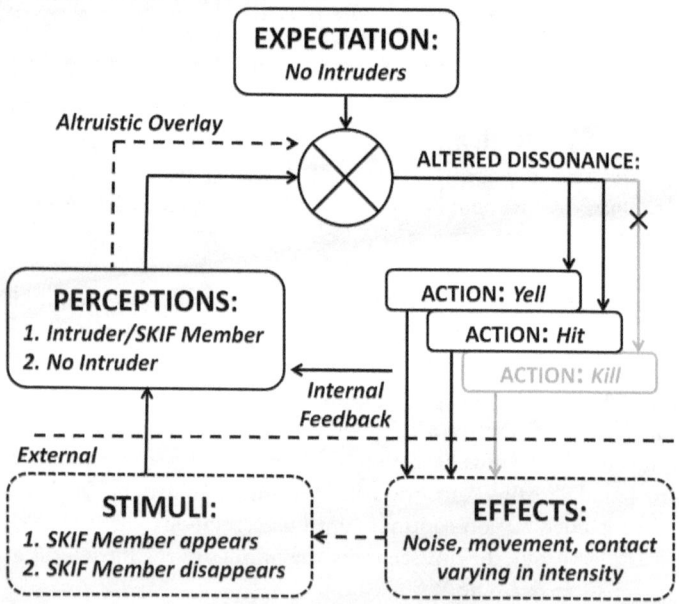

Figure 7.3. Continuation of hypothetical control mechanisms underlying hominid aggression and altruism (compare to Figure 7.2). This figure depicts aggression in an altruistic setting. The same dissonance pattern applies as in Figure 7.2, but in this case, the conflicting parties are members of the same SKIF group. This changes the outcome. The shared SKIF affiliation triggers an altruistic overlay that alters the dissonance computation and restricts the range of permissible action repertoires. As a result, the option to use lethal force is suppressed. Other aggressive actions may occur until perception matches expectation, but the *Kill* option is temporarily unavailable.

The notion that altruism arises simply by inhibitory neurons suppressing the neural activity that enables violent aggression may disappoint the true believers in altruism as a potent moral force in human life. Likewise, sociobiologists may dismiss it as simply an arbitrary mechanism that begs the question of altruism as an evolutionarily stable behavioral repertoire. No doubt, the schematics in Figures 7.2 and 7.3 grossly oversimplify real neural systems and functions. However, given the thesis of the present work, that control interactions underlie the unique behavioral repertoires of the human species, the fact that such simple mechanisms can capture the basic interplay of aggression and altruism is at least instructive. Perhaps more sophisticated control models will eventually bring all the relevant facts of these complex processes into a compelling, unified whole.

Homo beneficus and the Human Strategy

The inescapable conclusion of this analysis is that altruism and aggression are actually co-evolutionary traits, inextricably linked over the course of human evolution. Neither has dominated and neither has disappeared. This duality explains why, at times, we humans can be unconscionably harsh and cruel, while at other times, inexplicably kind and self-sacrificing. We are programmed for both. Our fate as a species declines if either trait dominates or disappears.

In the context of the control theory, if aggression is a tool to control events, then altruism is a tool to control aggression. Altruism emerged because the alternative, unchecked violence, pushed its practitioners into extinction. Decimated by genocidal conflict, burdened by low birthrates, and withered by infant mortality, such groups eventually had to succumb, no matter how dominant they might have been for some brief period. In contrast, those portions of the population that felt the subtle tug of altruism endured. While their altruistic behavior may have reduced their individual fitness, sometimes seriously, unrestrained aggression would have reduced it even more. The name *Homo beneficus* refers to those human variants that first acquired and expressed the various forms of altruism, and thereby escaped the slide into the oblivion threatened by *Homo bellicosus*.

Altruism underlies all aspects of human social behavior, constantly exerting its moderating effect on aggression. In reproduction, for example, we can imagine a society where procreation exclusively involves violent rape, where strength, cunning, and brutality are the primary determinants of reproductive success. Human evolution did not go that way because the altruism inherent in pair-bonding produced a better long-term result. Similarly, evolution could have produced less altricial offspring, freeing parents to compete more selfishly and aggressively. This did not happen because building big brains required more time and energy than building big muscles and bones. Pure aggression could have produced super hominids, standing eight feet tall with rhino-like strength and cheetah-like speed. But they would have had small brains, and might have eaten their young. Instead, nature built a big-brained, weak-muscled species that loves its children. For the moment, the latter appears the better solution.

The specter of genocide may yet prove that the human strategy lacks long-term adaptive value, that in our inexorable quest for control, we will eventually kill each other faster than we can reproduce. Nature, of course, will simply let us die, perhaps fossilizing our bones as evidence for future beings of an interesting, but short-lived, evolutionary sidetrack. We remain a species that can annihilate itself. As we move forward, the gentle tug of altruism may one of our few sustaining hopes.

CHAPTER 8

Homo humanitas, the Enculturated:
A Sea of Trivialities

Taming the SKIFs

If altruism were the only source of moderation against aggression, human society would be very different. Altruism operates best within SKIF communities where its moderating influences help prevent the escalation of unchecked aggression. Although competition still exists within SKIFs, it rarely degenerates into lethal violence, unless the group is extraordinarily dysfunctional or the competitive pressures are extraordinarily intense. Altruism is a social force that promotes this internal harmony.

Outside the SKIF community, however, it is a very different story. Altruism has considerably less impact between SKIFs, due primarily to lower levels of kinship, inlawship, and friendship. The result is an increase in xenophobic hostility, which can easily turn into lethal aggression. Between groups, violence remains an acceptable, even preferred, tactic for controlling events. Absent any additional social restraint, such aggression can degenerate into atrocity and genocide. Social systems caught in this xenophobic web remain independent and isolated, much as we see in present-day chimpanzee societies. We can envision the same pattern in the ancestral hominids: a population scattered in an archipelago of SKIFs, each island holding about one hundred to two hundred individuals (Dunbar 1996), and each regarding itself as the center of the universe.

Clearly, this pattern did not endure in human evolution. Modern human beings generally live in societies that greatly exceed Dunbar's 150-person benchmark. Some modern states number in the millions, equivalent to the fusion of thousands of SKIF communities. Such aggregations dwarf those of other primates, including our close genetic relatives, the chimpanzees and bonobos. The power of such numbers, applied cooperatively, gives the human species an enormous advantage in competition with other species.

Richerson and Boyd (1998) use the term *ultra-sociality* to describe the propensity of humans to aggregate into large communities. They suggest that cultural factors, combined with receptive genetic predispositions, account for this uniquely human social pattern. Thus, the emergence of human culture appears to mark an evolutionary path out of the quagmire of xenophobia. It somehow transformed human social systems from isolated competitive SKIFs into unified cooperatives. Something in the evolutionary development of human culture provided an antidote to the social poison of xenophobia. In the following, I shall elaborate on this theme, starting with some traditional views of culture, and then moving into more innovative characterizations.

Nature-Nurture Redux

Broadly defined, culture comprises the system of shared beliefs, values, customs, behaviors, and artifacts that social groups use to cope with their world and with each other, and which they transmit from generation to generation through learning (Bates and Plog 1990; Chapple 1970). Culture includes the totality of socially transmitted behavior patterns, arts, beliefs, institutions, and all other products of work and thought. Alfred Kroeber (1928) cited eight elements in his definition of culture:

1. *Innovation.* New behavior emerges through invention or modification.
2. *Dissemination.* Nongenetic transmission from innovator to others.
3. *Standardization.* Consistent, repeatable, sometimes stylized forms.
4. *Durability.* Continues in absence of innovator.
5. *Diffusion.* Spreads across groups.
6. *Tradition.* Endures across generations.
7. *Species-Specific.* Moves within species, rarely beyond.
8. *Transcendent.* Not directly determined by the immediate biophysical environment.

Patterns of behavior or knowledge that satisfy these criteria qualify as cultural phenomena, regardless of species, social organization, or evolutionary history.

The cultural components of human behavior are often contrasted with the deep-seated biological components we call *instincts*. Instincts are unlearned behavioral patterns, that is, actions that arise fully formed with no prior learning experiences or acquired memories. Their biological foundations lie closer to the neural and physiological structures encoded by the genes and elaborated during embryological development. Thus, certain repertoires, such as feeding, sexual arousal, or holding one's breath under water have powerful instinctive foundations. In contrast, behaviors such as writing books, playing video games, and driving cars are quite obviously learned.

The attempt to classify various human behaviors into cultural and instinctive categories has had a long and fractious history. It was the prime topic of the classic nature-nurture debates of the past century (Ridley 2003). For many years, biologists, sociologists, ethologists, psychologists and others argued at length as to the origins of human behavior, specifically whether it traced primarily to genetics (instinct) or learning (culture). This interplay produced many important discoveries, although political or social agendas sometimes tainted their interpretations. Depending on one's point of view, the opposing side was trying either to reduce humanity to the level of the animals, or to elevate humanity beyond the reach of biology. Needless to say, the great debate often created great confusion.

I will not attempt to recap the whole nature-nurture story here, except to state the now obvious conclusion: that a great deal of evidence exists for both nature and nurture in almost all human behavior. Far from being an all-or-none proposition, behavior often contains elements of both learning and instinct. Indeed, they sometimes blend so homogeneously that it is impossible to separate their respective contributions. For example, consider a human child just starting to walk upright. We routinely say that the child *learns* this skill and that the parent *teaches* it. But then, by the strict definition of culture as learned behavior, we would have to say that bipedalism is a hominid cultural tradition, rather than a genetic predisposition. Clearly, this stretches the point, as most anthropologists and biologists consider bipedalism to have strong genetic determinants. So which view holds, instinct or culture?

Consider another example: hunting behavior in young cats. Ethological studies suggest that a juvenile cat learns to hunt by observing its mother. Lack of exposure to a hunting mother usually dooms a young cat to a relatively unsuccessful career as a hunter. Hunting behavior obviously propagates across feline populations and down through the generations. So if it is learned behavior, then we should again argue that hunting in cats is a cultural tradition, rather than an instinct. Again, most biologists would disagree.

Social scientists now tend to agree that most behavioral repertoires emerge from the interaction of learning and instinct (Ridley 2003). In the case of hunting cats, for example, careful analysis suggests that the mother cat does not *teach* hunting to her offspring. Rather, her behavior *stimulates* or *shapes* the instinctive predisposition to hunt in the young cats, which subsequently consolidates into normal hunting behavior. Without the proper stimulation, the hunting instinct does not consolidate correctly, and the deprived cats mature into poor hunters. They might attempt to hunt, following their core instinct, but their success suffers from lack of shaping to an appropriate model.

Similarly, in human children just beginning to walk or talk, we presume that parents' *teaching* is basically just stimulating the developing child's underlying predispositions, which then consolidate into mature behavior. Bipedal locomotion and language surely have strong instinctive foundations, so it is likely that children would eventually attempt to walk and talk even if their parents did not participate. But absent parental stimulation, a child's unshaped locomotor and speech patterns may be abnormal.

The fact that behavior seldom emerges fully formed during development (instinct), that it requires environmental exposure to consolidate properly (learning), suggests that purely genetic factors have limits in mediating complex behavior. Recall that genes encode proteins not behaviors. Proteins in turn build and regulate the machinery and structures of cells and tissues. Only then does behavior emerge, typically many physiological steps away from the original genes. In some instances, a single variation in a single gene can trigger a specific behavioral change, but more often than not, genetic variations produce rather diffuse and wide-ranging behavioral effects. These genetic factors may predispose an individual to behave in certain ways, so that when the appropriate environmental situations arise, their underlying neural structures bias them toward particular actions. But the simple assertion that genes *produce* behavior demands considerable caution. In most cases, genetic variations cast only an indirect shadow on specific behavioral repertoires.

Once human children pass through the critical periods for shaping instincts, their behavior transitions to an adult pattern of cultural learning. In this mode, the human brain seems to acquire knowledge almost effortlessly. Human children assimilate huge amounts of information about their world. Many theorists see this stage of human development as the primary stronghold of the *tabula rasa* or *blank slate* school. In this view, cultural knowledge passes to children and young adults like words filling the empty pages of a book. While instincts may still bias the child's behavior in certain ways, most cultural repertoires do not spring instinctively into existence. For example, we are born with an instinctive predisposition for learning languages, but we are born not knowing any specific language. Similarly, we may have an instinctive predisposition toward technological manipulations, but we are born not knowing how to make fire or fashion cutting tools or select and prepare food. In these areas, the notion of a blank slate on which cultural specifics are written makes some sense. During this phase, children assimilate and consolidate the specific cultural patterns that give them the skills and knowledge to function successfully as adults in their society. At the same time, of course, they also acquire all the biases and misinformation of the culture, and thereby perpetuate the same social problems and weaknesses to the next generation.

For many years, sociologists saw culture as exclusively human. While there no doubt that human cultural development greatly exceeds that of any other species, extensive fieldwork over the past half-century has revealed clear evidence of cultural transmission of behavior and knowledge in nonhuman species, particularly the primates (Imanishi 1957; Whiten et al. 1999; Goodall 1986; Nishida 1990; Boesch and Boesch-Aschermann 2000; Wrangham et al. 1994). In the chimpanzee studies, researchers have cataloged thirty-nine cultural behaviors, plus twenty-six additional candidates (Whiten and Boesch 2001). I mentioned a few of these in Chapter 5, such as nut-cracking and insect fishing. These repertoires fit the cultural requirements outlined above as much as any human cultural activity, so we cannot say that cultural development is a unique human characteristic. However, we can argue that its quantitative expressions in the human species represent a qualitative shift in its social importance.

The discovery of nonhuman culture has reawakened interest in the biological basis of human culture and its evolutionary history. If our closest primate relatives display longstanding cultural traditions, we naturally ask whether our most ancient ancestors also had the same propensities. And if so, what adaptive outcomes triggered the massive expansion of culture in the hominid line, as compared to the more limited repertoires of our genetic cousins. Some thoughts along these lines follow.

Memes and Genes

The classic nature-nurture debate has its modern mechanistic equivalent in the interplay of memes and genes. The term *meme* originated with Richard

Dawkins in *The Selfish Gene* in 1976 as a corollary to his *gene's eye-view* conceptualization. In genetics, Dawkins suggested that the mechanisms of natural selection always produce outcomes in which gene seemingly acts with selfish intent. Thus, a genetic variant that improves reproductive fitness tends to generate more copies of itself. In effect, the gene makes the organism more successful, so the organism can in turn make the gene more plentiful in the population, thus fulfilling its selfish mission. Of course, genes have no intent, selfish or otherwise, but the mathematics of natural selection allows such animistic interpretations.

Dawkins generalized this concept to show that any selectionistic process supports the same self-centered interpretation. As a corollary to selfish genes, he posited the concept of selfish *memes* as the basis for cultural development. In Dawkins's view, memes are simply behavioral propensities that replicate and propagate through a society by imitation and repetition. Such elements might include movements, vocalizations, decorations, displays, and so on. People observe these patterns and choose either to imitate them or to ignore them. Patterns that are imitated (selected) are said to move into new carriers (replication) who in turn exhibit the same behavior, which again are imitated by other individuals, and so on. In this way, the original pattern propagates across the population and down through the generations. Memes that fail to inspire imitation disappear (extinction), while those that are repeatedly imitated survive. Theorists coined the term *memetics* as the study of memes. Many authors have expanded and clarified the concept (Blackmore 1999; Aunger 2000, 2002).

As an information theory, memetics deals with the replication and propagation of information patterns that have adaptive meanings to both sender and receiver. Like Gregor Mendel's original articulation of the gene, enunciated long before science had any knowledge of actual gene structure, the meme concept provides an aura of tangibility to an otherwise nebulous phenomenon. It gives us a convenient mechanistic way of talking about a vast and complicated subject. The antecedents of meme theory came from F. T. Cloak (1975), who differentiated between the set of cultural instructions people carry in their nervous systems, which he called i-culture, and the material structures and physical outputs of behaviors specified by those instructions, which he called m-culture. Artifacts fall into the m-culture category, as do social organizations, conversations, technologies, ideologies, behavioral displays, and so on. Dawkins folded Cloak's concepts into a selectionistic framework that showed how i-cultures comprise an assemblage of memes whose expression at the m-cultural level determines their selective value. Thus, a meme encodes the unit of information residing covertly in the brain. Its activity produces an overt object or event that observers can experience and imitate.

Imitation is crucial to any theory of memetics (Dawkins 1976; Blackmore 1999). An individual behaves in some particular way in a given setting, and witnesses, for whatever reason, elect to perform the same behavior in comparable settings. For example, suppose an individual frightens away a predator by banging two rocks together. A witness to the event remembers it and displays the same behavior in subsequent encounters with predators. Others observe and imitate this action, and eventually the behavior spreads throughout the group. Thus, the original meme, the idea of banging rocks together to deter predation, may propagate from one desperate innovator throughout the local culture. Viewing this progression from the outside, an observer concludes that a meme residing in the first individual (sender) replicates and propagates to the others (receivers). This process may repeat any number of times.

Biologists and behaviorists have long recognized imitation as the primary genetic predisposition that facilitates learning in many species, particularly the primates. Premack and Premack (2003) cite imitation as a fundamental enabler in the emergence of intelligent behavior, both in humans and our genetic cousins. As such, imitation plays a central role in human cultural development.

With each transmission, memes may be modified slightly. A given meme exists in a world of countless others, all residing and interacting in individual brains. As any given meme transits through the community, its context varies with the changing cognitive environments of its hosts. During replication and propagation, hosts may alter the meme's content, either deliberately or inadvertently. In this way, variations (mutations) of the original meme appear, each depicting a different version of the original event. Over many cycles, such variations may drift in form and content until, in the extreme, they barely resemble the original. As variations proliferate, receivers must choose among the versions. Such individual selections determine the fate of each variation.

In genetics, selection failure often takes the form of death without reproduction. In memetics, the process is usually less consequential, although in certain circumstances, the wrong choice can bring sudden death, as in our example above of repelling predators. In any challenging environment, the members of a social group must each acquire and share essential survival and reproductive skills and, in turn, must propagate them to their offspring. In advanced species, learning plays a critical role in this process. Each generation remembers the lessons of the previous one, and teaches the next. Natural selection rewards those genetic lines that best accomplish this process of cultural transmission and retention.

Trivialities

Now, if everything we learned from our parents and peers were life-critical, the evolution of culture might be easy to explain. Individuals that acquire the capacity to learn, remember, and teach are better able to survive life's critical challenges and to pass that knowledge on to their progeny. Individuals deficient in some aspect of enculturation, who either cannot learn or cannot teach, face a greater likelihood of extinction. Thus, genetic predispositions for imitating, learning, and retaining life-critical information have clear adaptive value for any species trying to survive in a challenging environment. The human species has invested heavily in and benefited greatly from its strong predisposition for cultural learning.

In some ways, however, human culture is paradoxical. Many, if not most, human cultural activities and expressions hardly seem life-critical. Indeed, most of human culture appears disturbingly trivial. Consider, for example, clothing and hair styles, tattoos and personal adornments, decoration of dwellings and tools, language idioms, superstitions and taboos, pop rock, Barbie dolls, and thousands of other modern fads and vanities. One is hard-pressed to relate the proliferation of such seemingly trivial displays of human culture, both modern and ancient, directly to the biological imperatives of adaptivity, survival, and reproductive success. Tattoos do not repel predators or prevent disease. Tokens and fetishes attached to tools and weapons do not increase the gatherer's yield or the hunter's success. Speaking slang does not guarantee a fertile mate or healthy children. Yet, in any given human social group, such trivialities abound. They are faithfully imitated, embellished, and propagated throughout the local culture and down through the generations, often with great expense of energy, time, and materials. How do we explain this? What biological imperative do such trivialities serve? More importantly, with the enormous investment in brain capacity in genus *Homo*, why is so much of it seemingly occupied with cultural rubbish, at least from the evolutionary perspective? Such questions pose a fundamental challenge for memetics, indeed for cultural models in general. The following section offers one possible explanation.

Cultural Signaling and the Control of Xenophobia

A complex social environment bombards an individual with a multitude of would-be memes. Some will evoke imitation and perpetuate, others will be lost in the noise. To explain the emergence and properties of human culture, meme theory must explain how and why individuals select certain memes and ignore others. Which behaviors, among the many we observe, do we decide to imitate? And why, among the many we choose to imitate, are so many of no apparent significance to our immediate biological survival?

To answer, we need to divide human culture into two major components: 1) cultural diffusion *within* the traditional SKIF community; and 2) cultural signaling *between* SKIF communities.

Let's look at each of these in more detail.

The first is the more powerful and the more evolutionarily defensible. Human children usually grow up in the supportive environment of their SKIF community, comprising parents, siblings, relatives, inlaws, and friends. Experiences with these individuals dominate early human social life. In this nurturing environment, children learn by imitating those closest to them. They acquire the local SKIF culture, its language, customs, rules, technologies, and territory. They recognize the names and faces of group members. In time, they discover their place in the dominance-subordinance hierarchy, sometimes by unpleasant competition, but seldom by life-threatening violence. In short, the major source of cultural programming in childhood comes from within the SKIF community, defining an individual's lifelong expectations and preferences. At this stage, the child rarely chooses which memes to accept and which to ignore. Proximity and the predisposition to imitate guarantee the acceptance of most memes, which assures the perpetuation of the local SKIF culture, for better or worse.

As children mature and take their places in the local SKIF community, they begin to acquire the typical xenophobic posture toward rival SKIFs. This is an important transformation. Young children generally exhibit little xenophobia. In theory, they would thrive equally well in one SKIF or the other, given the same parental attention and social stimulation. Older children, however, adopt and display the socio-political attributes of their home group, eventually expressing the characteristic wariness and hostility toward other groups. The consequence of this shift is the perpetuation of xenophobic exclusivity between groups, both genetic and cultural. Each SKIF community tends to maintain its own gene pool and its own cultural profile.

In a xenophobic environment, memes typically do not flow any better than genes. Groups that occasionally exchange reproductive partners as a way of insuring genetic diversity may have some collateral opportunity for cultural exchange. The émigrés carry the memes of their home group. Their adoptive group may imitate some of these, thereby adding new cultural elements. In other cases, the immigrants may have to abandon their previous cultural preferences in order to gain acceptance into the new group. Thus, reproductive exchanges between SKIFs may promote meme propagation in some circumstances, but be actively resisted in others.

In the absence of robust cultural exchanges between groups, the cultural potential of any one group is limited to its own narrow intellectual and physical resources. A given group can go no further than the talent of its membership. Innovations in technology, improvements in basic skills, healthier behaviors, all must originate within the group. Blessed with intelligent innovators, some groups may gain strength. Others, lacking the requisite talent, may lag. As the innovative groups develop better solutions to problems, such as better weapons, efficient transportation, improved diet, and so on, they may gain strength relative to their less capable competitors. Such cultural innovations may fuel successful aggression against the less innovative groups. The losers in this xenophobic struggle face banishment, slavery, or death. Thus, in the territorial struggle among human groups, cultural innovation and knowledge represents a clear competitive advantage.

In theory, successful innovative cultures could conquer neighboring groups and dominate a steadily widening expanse of territory. However, the natural limits of SKIF communities make this problematic. If Robin Dunbar (1996) is correct, a SKIF community, no matter how innovative or culturally enlightened, seldom supports a population of more than 150 members on the average. Larger groups tend to split into subgroups, or factions. These factions may pull away from one another into separate ranges, eventually perhaps adopting the xenophobic stance of competing groups. The fission of a larger SKIF produces two smaller competitors who start with the same cultural repertoires. While each subgroup may be stronger than the surrounding groups, they start at approximately the same point relative to one another. In that competition, they face the possibility of prolonged stalemate, perhaps escalating eventually into perpetual conflict. As we outlined previously in Chapter 6, such extended conflicts represent a burden than potentially weakens both sides.

Faced with xenophobic stalemate, competing human societies must eventually come to a critical decision point: (1) continue their competition, with the ever-present threat of withering violence; or (2) moderate their xenophobia in favor of more cooperation, albeit with less separation and independence. As we outlined in Chapter 6 above, the former represents a dangerous path toward extinction. However, the latter appears to violate basic human nature, at least in its prototypical forms. Peaceful coexistence among competing SKIFs clearly demands some new evolutionary mechanism or trait that moderates the inherent xenophobia between groups. This shift constitutes, I suggest, a unique human development: *cultural signaling*.

Cultural signaling has two distinct but closely related components. First, the signaling processes that occur *within* SKIFs, and second, those that involve displays *between* SKIFs. Let's look at how these work.

The cultural signaling that occurs within SKIF communities is at once familiar and paradoxical. Every established group of humans tends to have a particular set of customs and traditions that distinguishes it from other groups. These include unique patterns of personal adornment, decoration, design, technology, behavioral rituals, language dialects, and so on. We take such cultural identities for granted. Ask why we adopt particular customs or displays, we say simply "It is our way," as if that explains it. Such patterns have been the fundamental research material of anthropologists for many decades.

What is paradoxical about such within-group cultural displays is their seeming irrelevance to the successful functioning of the group. For example, suppose a particular group has the cultural tradition of wearing white hats. Every individual learns and adopts this display. But what actually is its value in the SKIF's internal social system? How does it contribute to the success of the group? The common answer is that it somehow binds the social group together, but this makes very little sense. As we discussed in Chapters 6 and 7, the members of a SKIF are held together by powerful forces of pair-bonding, kinship, inlawship, and friendship. From the biological perspective, these are the true binding forces within a SKIF community, not tattoos, adornments, or clothing choices. Thus, two individuals support one another because they are brothers, cousins, inlaws, or friends, not because they both wear white hats. It is difficult to see how trivialities, such as clothing choices, could have enduring impacts on the evolutionary underpinnings of the SKIF social system.

Why then do such within-group displays appear ubiquitously in human SKIFs? One possible answer is that they are actually part of the xenophobic social repertoire. That is, they are not reflections of the internal social dynamics *within* the group, but rather internal echoes of the external social dynamics operating *between* groups. In other words, two individuals both wear white hats not because it makes them better social partners, but because it generates a perceivable manifestation of their shared membership in one particular social group *as opposed to* all other groups. Wearing a white hat is not an internally directed display; it is an external one. Its primary purpose is to alert, perhaps even to warn, competitive SKIF members in the neighborhood that they are encroaching on another SKIF's domain. By adopting such displays across the entire membership, one SKIF intensifies its perceived separation from other groups, and amplifies its observable presence to those other groups. The effect of such cultural displays, at least in the early stages of enculturation, is to reinforce the group's natural xenophobia. Thus, when a particular SKIF community says "It is our way," what they mean is, "It is not *their* way," referring to their local competitors.

This kind of internal xenophobic echo accords with the general hypothesis put forward in Chapter 6. There I suggested that the binding forces in human SKIF communities were much stronger than those of other social apes, like the chimpanzees. The stronger human propensities for pair-bonding, parental investment in offspring, and kinship recognition, both direct and indirect, make human SKIFs much more cohesive and unified than comparable nonhuman societies, and therefore potentially much more competitive with one another. Their adoption of distinctive cultural displays to augment this competitive stance should not be surprising. Their shared customs are simply visible manifestations of their already recognized separateness from other groups.

Clearly, however, the strong xenophobia exhibited by early hominid SKIFs must have given way at some point in human evolution to a more moderate, cooperative stance. Otherwise, the ultra-sociality that characterizes modern human societies would never have emerged. How did the ancient SKIFs, who engaged in such elaborate efforts to preserve and display their mutual xenophobia, ever come together in a larger social system?

The answer to this question opens yet another cultural paradox. It is simply that the same kinds of cultural displays created originally to emphasize xenophobic separation can, when modified slightly, signal the exact opposite, that is, a moderation of xenophobic hostility. Once the competing SKIFs began using cultural displays to deter or taunt one another, they were in fact *communicating* with one another, using cultural signals. Those first messages were, of course, relatively hostile. But the groups were paying attention to one another, rather than feigning indifference or blindly attacking. Once they started observing one another, these competitors gained a potential channel to signal one another in a less xenophobic context. The simplest way to do this was memetic: each group began imitating selected cultural signals they observed in the other group. Figure 8.1 illustrates an example of this process.

The figure shows two human SKIF groups. Following their within-group traditions, Group *A* wears white hats and white vests, while group *B* wears black hats and black vests. Let's assume that historically these groups have maintained a xenophobic posture toward one another, aggressively competing for resources, perhaps to the point of lethal violence. Let's suppose also that in this case, the result is stalemate. Neither side is able to defeat the other. Both sides, however, continue to experience population growth and resource pressures, which inevitably forces them into closer contact. Now, let's suppose that in the gene pools of these two competing groups, there are genetic variations that predispose the respective populations to expand their cultural

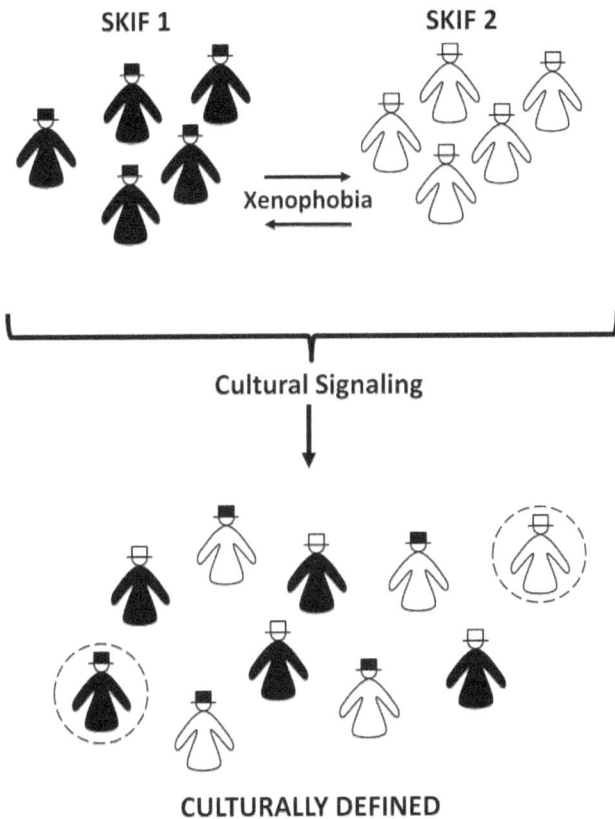

SKIF 1   SKIF 2

Xenophobia

Cultural Signaling

CULTURALLY DEFINED
COOPERATIVE

Figure 8.1. Cultural signaling reduces xenophobic aggression. (Top) Two SKIF groups adopt a traditional xenophobic posture. Each group retains its exclusive within-group cultural attributes (e.g., like-colored hats and coats). (Bottom) The emergence of cultural signaling promotes the merger of the competitive groups into a larger, culturally-defined cooperative. Each group adopts selected elements of the other's cultural display (e.g., *white* hats and *black* coats). This masks their original cultural profile, which reduces the potential for xenophobic provocation. The original SKIF groups still exist within the larger cooperative, but they are no longer as visible. Competitive groups may now intermingle with less chance of aggression. Some individuals (circled) may opt to retain their traditional cultural displays with their attendant risks of xenophobic provocation. Cultural signaling helps to explain why human societies can grow so large, and why so much cultural activity seems biologically and evolutionarily trivial.

displays. For example, certain members of group *A* may relax their strict white-white dress code and start wearing black vests while retaining their traditional white hats. Members of Group *B* might undergo a similar cultural relaxation, donning white vests but retaining their traditional black hats. Thus, each group begins to display hybrid cultural patterns that a) partially mask their own cultural tradition; and b) partially match the cultural tradition of their competitor.

These novel patterns, neither white-white nor black-black, are culturally ambiguous. They may therefore represent weaker triggers for traditional xenophobic reactions. Where a white-white pattern would have provoked hostility from a black-black observer, the white-black hybrid has less effect. Practiced reciprocally by the two groups, the net effect of such nontraditional cultural displays is a temporary disruption of the normal xenophobic pattern.

What matters here is not the form of the novel cultural signals, but the reciprocal exchange itself. Clothing colors and patterns are trivialities, without enduring biological importance. However, as social signals, they represent potential invitations to cooperate. More importantly, they offer a mechanism for the evolving human social system to finesse its ancient *us-versus-them* xenophobia. Using this new signaling mode, social competitors opened the door to a new type of social aggregate, a loose affiliation of previously hostile SKIFs. Over time, these aggregations opened the door to reciprocal exchanges of information, technology, and resources among the groups, thus establishing the first tentative characteristics of human ultra-sociality.

Like other cultural activities, the engine that drives such cultural signaling is memetic: the propensity to observe, imitate, innovate, and propagate certain behaviors. What made cultural signaling evolutionarily important, and unlike other cultural processes, was that it operated *between* groups. Its adaptive value was the moderation of xenophobia, which led in turn to a reduction in aggressive competition, resource depletion, and death. By selecting and imitating each other's cultural displays, competing groups found a communication channel that led to relative peace. This clearly marked the evolutionary path to *Homo humanitas*.

Once they stopped fighting and started cooperating, hominid groups realized immediate economic and social benefits. Cultural signaling allowed humans to extend their social contacts to a wider circle, albeit with less intimate relationships. The archipelago of isolated SKIFs became a contiguous social network through which resources, knowledge, technologies, and reproductive partners could flow more freely. No longer did every group have to go it alone, each reinventing the same technologies and processes, each forced to harvest its resources with limited hands. By observing and imitating other groups, individual SKIFs benefited from the best inventions and ideas from the entire accessible population. The resulting proliferation of shared technologies and process improvements increased the power of each group, and therefore benefited each individual.

As cultural signaling evolved, the original SKIF groups did not disappear. Even today, they remain embedded in our ultra-social human communities. They still comprise extended families, clans, coalitions, alliances, and friendships, much as in ancient times when they occupied exclusive territories and defended them to the death. While modern humans still owe allegiance to their SKIF group, and will defend it aggressively if threatened, they also recognize the benefit of cultural linkages to other groups that enable cooperation within a larger community.

Just as SKIFs did not disappear from ultra-social human systems, xenophobia did not disappear from the behavioral repertoires of any group within the extended social system. Breakdowns in cultural signaling, such as factions reverting to their original provocative displays, may have periodically disrupted the new cooperatives. In the extreme, such breakdowns may have reignited historic patterns of lethal violence, as if there had been no intervening phase of cooperation. Even today, the breakdown of cultural signaling regularly triggers xenophobic outbursts in the forms of racism, nationalism, and religious intolerance. Such hostility may even reach genocidal intensity, unless or until the conflicting groups can find some acceptable new set of signals that restores mutual acceptance and cooperation.

The emergence of between-group signaling represents an important evolutionary turning point in the history of the hominids. It accentuated their divergence from their ape cousins. While both the early apes and the hominids probably acquired considerable cultural knowledge within their traditional SKIF circles, only the emerging humans found a way to transmit such knowledge across the xenophobic divide. The chimpanzees and bonobos still cling to their historic internal focus, showing little or no interest in or ability to learn from other groups.

The ancestral hominids who discovered cultural signaling added a new dimension to their social systems. In essence, cultural signaling becomes the sixth building block in human social systems, bridging the gap between xenophobia and the social elements that operate entirely within the SKIFs. This addition is shown in Figure 8.2. On this new middle ground, *Homo humanitas* built a huge diversity of culturally defined communities of interest. The cooperative power of these loose aggregations far exceeded the economic, technological, and ecological power of individual groups. With this singular social innovation, the emerging hominids gained a selective advantage over the other apes, an advantage they have parlayed into a crushing evolutionary force.

In the previous chapter, I suggested a mechanistic control model of altruism. The same kind of mechanisms can help explain the neural foundations of cultural signaling.

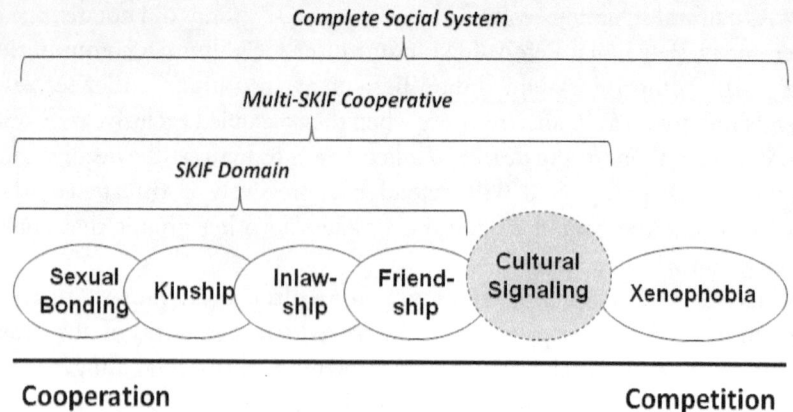

Figure 8.2. Cultural signaling adds a uniquely human sixth building block to the social map. Compare with Figures 6.1 and 7.1. Cultural signaling bridges the gap between xenophobic separation and the various forms of within-group cooperation. It represents a new class of social interactions that are less cooperative than traditional SKIF interactions, but also less hostile than traditional xenophobic interactions.

The Control Interpretation of Cultural Signaling

The control interpretation of cultural signaling rests on the expansion and proliferation of expectations, both within-group and between-group. As the hominid brain grew steadily larger, the additional capacity permitted much greater numbers of reference expectations, often in multiple forms each slightly modified from the original. Recall that memes often undergo transformation and variation as they propagate from one brain to another. The greater the brain's carrying capacity for memetic patterns, the more such patterns any one brain can contain. Depending on how the brain acquires and codes such representations in memory, any or all of such memetic patterns may serve as reference expectations at any given time. This represents the critical evolutionary transition from the pre-*humanitas* hominids to the *humanitas* form. Figures 8.3 and 8.4 illustrate this process.

Figure 8.3 shows a pre-*humanitas* structure in which the expectation deck includes only those cultural elements tightly held within an isolated SKIF community. In this example, the only available reference expectation regarding clothing displays is the traditional *black-black* pattern characteristic of the individual's home SKIF. It defines the one accepted clothing pattern recognized by all. Any perception of clothing patterns other than *black-black* must increase internal dissonance, which in turn must trigger action to change or eliminate the offending perception. In the cultural history of this particular SKIF, this exclusive *black-black* pattern is the established tradition.

In the neighboring territory, also occupied by a pre-*humanitas* SKIF, the same process has presumably occurred over the generations, but here it is the *white-white* preference that has become established. Whenever these two groups encounter one another, their mismatched perceptions and expectations increase their levels of dissonance and the probability of aggressive action to resolve the mismatch becomes more likely. This xenophobic posture has presumably existed for generations.

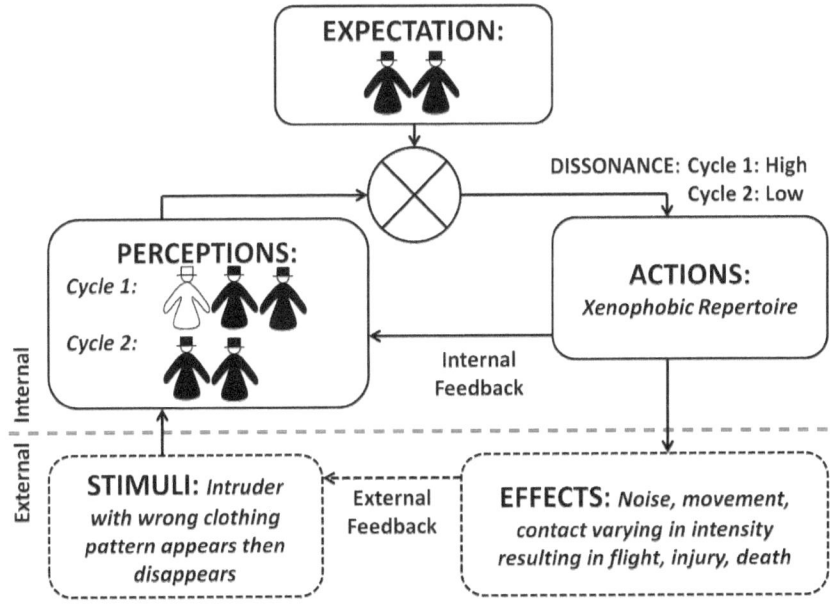

Figure 8.3. A mechanistic interpretation of cultural signaling (see also figure 8.4). This figure depicts the pre-*humanitas* hominids, where the primary social expectation is limited to one particular cultural pattern. Any departure from the traditional cultural display triggers a xenophobic behavioral response.

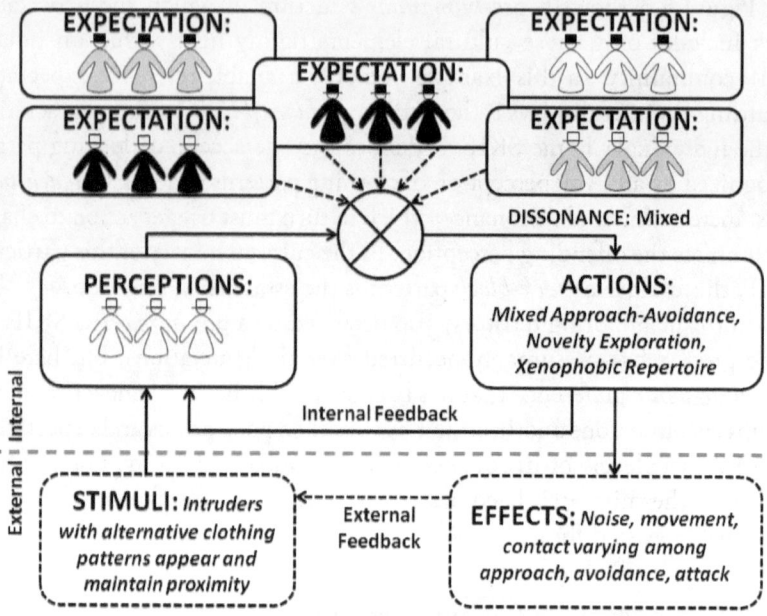

Figure 8.4. Continuation of the mechanistic interpretation of cultural signaling (see also figure 8.3). This figure depicts the evolution of *Homo humanitas*, which underwent an expansion and diversification of expectations, compared to the pre-*humanitas* form (Figure 8.3). With the proliferation of allowable social expectations, interactions among social groups generate fewer xenophobic provocations, and therefore reduce the potential for aggressive or violent behavior.

Figure 8.4 illustrates the hypothetical transformation to *Homo humanitas*. Here, the deck of reference expectations residing in each individual brain has grown somewhat larger by the addition of multiple variations of the original expectation. Thus, while the traditional *black-black* pattern is still present, several other versions, each slightly modified, now accompany the original in the larger brain. Figure 8.4 depicts these variations as *black-grey*, *grey-white*, and *black-white*. The individuals that carry these alternative expectations have potentially greater tolerance of clothing combinations that do not conform to the traditional *black-black* model. As they perceive other individuals displaying nontraditional clothing combinations, both within their own SKIF and across the competitive divide to other SKIFs, they are less likely to process it as a complete mismatch, and therefore less likely to experience high levels of dissonance. As a result, this variant is less inclined toward the full xenophobic repertoire when confronted by nontraditional clothing combinations. In effect, they display greater tolerance both of their within-group peers who depart from the traditional *black-black* mode of dress, and of their between-group competitors who depart from the traditional *white-white* mode. Thus, by the simple neurogenetic expansion of brain tissue, thereby creating the capacity to carry additional variations in expected cultural displays, the emerging humans become less likely to lapse into a full-blown xenophobic repertoire when they encounter any particular cultural display. This inherent restraint represents a clear departure from their pre-*humanitas* hominids and the other great apes.

Figures 8.3 and 8.4 undoubtedly oversimplify the actual mechanisms of cultural signaling in the human species. Real cultural settings involve an enormous multitude of perceptions, each compared to a vast range of individual expectations, leading to complex, often conflicting, dissonance and action. The mechanisms depicted here must be multiplied many millionfold and juxtaposed against one another. Like other human traits outlined in this book, the mechanistic interpretation of cultural signaling is simply a starting point for understanding how simple control systems might underlie complex cultural development.

Lies, Deceptions, and Viral Memes

While cultural signaling opened the door to peaceful coexistence among otherwise competing human groups, it was hardly a panacea. True, it allowed social aggregations to mask, if not truly moderate, their traditional xenophobia. It promoted social cooperation on a vastly larger scale than in other ape species, enabling participants to benefit from the exchange of knowledge and technologies, to exploit the resources of their combined territories, and to gain access to a more diversified pool of reproductive partners. However, cultural signaling has limits. The human species measures all actions and events against their potential for control, not cooperation. However beneficial cooperation may seem to society as a whole, it endures only because it conveys a greater sense of control to the individual participants. If any participant loses that sense of control, cooperation immediately breaks down and the social posture quickly reverts to xenophobia, competition, and perhaps lethal violence. Cultural signaling simply provides a communication channel that indicates the *intent* to cooperate. If the new arrangement does not promote a mutual sense of control among the individual participants, then the culturally defined social structure quickly degenerates.

Cultural signaling also introduces a new problem in social interactions: deception. Deception dwells in the vast middle ground between altruistic cooperation and lethal aggression. It attempts to maximize, covertly, the selfish agenda of the deceiver. In its simplest terms, a deception involves cultural signaling by the deceiver in order to evoke desired actions by the deceived. In most cases, the purpose of the deception is to achieve an outcome not readily attainable through honest cooperation.

While it is tempting to characterize deception as a cultural problem or weakness, it actually represents a logical consequence of our social evolution. The predisposition to control is the central driver of human activity, but it carries the potential for violent aggression if practiced unchecked. The moderating forces of altruism and cultural signaling may reduce this threat, but they do not necessarily guarantee optimal outcomes for all participants in the society. The logical resolution of these conflicting pressures is an alternative social strategy, deception, which offers the potential for selfish gain without the attendant risk of lethal aggression. Put simply, if we did not practice the art of deception, we would surely practice the art of murder.

That deception is a natural part of human interactions is quite evident (Smith 2004), but nonhuman primates also display the capability, as demonstrated by Richard Byrne and others. As we noted above, chimpanzees and other primates show evidence of cultural development, although quite limited compared to humans (Whiten et al. 1999). Byrne showed that these species also practice various forms of deception in their interactions with peers and competitors (Whiten and Byrne 1988, 1997). Moreover, he suggests that the propensity to deceive is directly correlated with the size of the brain's neocortex (Byrne and Corp 2004), precisely the brain structure implicated in the propensity for control.

The one saving grace for the human species with respect to deception is that it appears somewhat harder to lie than to tell the truth. We know that people vary greatly in their propensity to behave deceptively. Every society has its saints who display an excess of truth, and its sociopaths who are seemingly compelled to lie at every opportunity. That these behavioral propensities have a deeper physiological basis is suggested by modern polygraph (lie-detector) technology. Most people exhibit autonomic reflexes that manifest their internal dissonance when perpetrating a deception. This suggests that normal people have a natural resistance to deception, although it seemingly does not prevent them from lying when the chance of detection is remote. That sociopaths can apparently defeat polygraph testing argues that they are indeed physiological variants.

Meme theory offers an interesting twist on the problem of deception (Dawkins 1976; Blackmore 1999, Dennett 1995; Brodie 1995; Lynch 1996).

This is the concept of the *viral meme*. It refers to memes that deceptively promote their own replication and propagation, hence the comparison to viral pathogens. Chain letters are the classic example. A chain letter's primary content relates to making copies of itself and forwarding them to additional recipients. Ordinarily, chain letters include some inducement, such as good luck or popularity. Some request the receiver to make a financial investment on behalf of the previous sender with the inducement that, as senders themselves, they too will receive the benefit of such investments by future receivers. Of course, that rarely happens. Individuals who lose money seldom participate a second time, so the popularity of chain letters remains limited to the inexperienced and the chronically gullible.

The defense against viral memes lies in focusing on the primary adaptive value of meme propagation in normal situations. As we outlined above, individuals respond to memes in direct proportion to their value in solving control problems. The most effective viral memes typically address some important problem or desire. To endure, however, they must eventually deliver some tangible value. That is, actual perception must eventually match expectation. Failure to deliver on the promised control leads quickly to the meme's deletion. Perceptive individuals can usually assess promised solutions against realistic expectations. Those who fall for such confidence schemes repeatedly, or who risk significant value, often do so out of desperation, greed, or some perceptual deficiency.

Cultural signaling clearly has both positive and negative impacts. On the plus side, it gave our ancestors a way to escape unrelenting violence over territory, resources, and reproductive partners. It allowed competing groups to form unimaginably large cooperatives, concentrating force, technology, and knowledge. The downside lay in the misuse of cultural signals to defraud the unwary and to promote undeserved dominance among the few. Ultimately, cultural signaling is simply a social mechanism to help individuals control events in their lives. If general population fails to remain vigilant against the misuse of such tools to rob them of control, then *Homo humanitas* will almost tumble into extinction.

Homo humanitas and the Human Strategy

In ancestral times, cultural learning represented a powerful way by which children acquired knowledge about the world from their parents, grandparents, and, extending back, from their ancient ancestors. It gave them the tools they needed to function in their social milieu and a sense of place within a supportive group. Ancient hominid populations thrived in the steadily increasing knowledge and capabilities that cultural learning conveyed within their groups.

But these powerful cultural identities also separated groups from one another and intensified their traditional xenophobia. Left unchecked, cultural separation would have accelerated our ancestral tumble into withering violence. Only through the evolutionary innovations of *Homo humanitas* did the emerging human species escape this ancient xenophobic trap. In *humanitas*, there emerged a second form of culture, a signaling layer that enabled ancient clans to stop killing one another and start observing, imitating, and cooperating. This genetic variation, first carried by *Homo humanitas*, encoded a predisposition to invent, display, and perceive novel behavioral signals. These were often trivial and meaningless, but they signaled the intent to cooperate rather than compete. The widespread adoption and acceptance of such signals led to the intermingling of groups and the formation of much larger social structures. Once *Homo humanitas* escaped the xenophobic quagmire of *Homo bellicosus*, it took a giant step toward power among the earth's species, that is, toward becoming *Homo dominus*.

Modern human populations are still composed of the smaller social systems we call SKIFs. We continue to recognize our families, relatives, inlaws, and friends. We grant them special status in our lives. When threatened, we will defend them aggressively, even violently if necessary. In this, we are not much different from those ancient hominid ancestors who first moved toward the human path. Where we differ is in the vast elaboration of other cultural signals that allow us to intermingle with other SKIF communities in huge numbers. Where the ancient SKIFs might have fought one another on a daily basis, the emerging *humanitas* groups discovered the selective advantages of peaceful cooperation, which ultimately outweighed the benefits of conflict.

While *Homo humanitas* adopted cultural signaling for an enormously important purpose, its overt manifestations often seem trivial, either delightfully or disturbingly, from the perspective of human evolution. Social science struggles with this seeming paradox: that something so crucial to our success takes the form of such trivial objects and events. *Homo humanitas* had the answer. The adaptive value of cultural signals lies not in the objects and actions themselves but in the very act of communicating. Combined with the life-critical learning and innovation *within* SKIF groups, the emergence of stable communication channels *between* groups triggered an explosion of knowledge, technology, and cooperation that propelled *Homo humanitas* toward modern ultra-sociality. While it did not eliminate the problems of deception, warfare, and genocide, it gave humans a social system that pushed xenophobic hostility toward the periphery of life. Our ape cousins never found this path, and they were left behind.

CHAPTER 9

Homo aestheticus, the Artist: The Art of Losing Control

The Problem of Art

There is an aspect of human culture that deserves special attention: the arts. We traditionally define the arts as those forms of culture encompassing the ancient activities of song, chant, percussion, instrumental music, dance, drama, costumes, decoration, painting, sculpture, architecture, and writing. Modern additions include photography, motion pictures, radio, television, and so on. I refer to the arts as a *problem* not because they do not fit the characterization of culture outlined in Chapter 8, but rather because, practiced to their extremes, human artistic expressions go so far beyond our simple notions of cultural learning and signaling that it seems as if something more must be going on. While most members of a social group can learn the survival skills necessary to cope with their environment, such as harnessing energy or gathering food, and nearly all can master the group's established customs and traditions in dress, gestures, speech, and so on, very few members of any group ever acquire the skills of the true artist. The ability to create unique objects and memorable displays requires talent that goes beyond most cultural norms. Applying the term *culture* both to the mundane activities of daily life, like styling one's hair, and to extraordinary artistic achievements, like the cave paintings at Chauvet or Michelangelo's *David*, seems to miss a fundamental point. The rarity of true artists and the transcendent qualities of their work appear to belong on a different branch of the cultural tree. Perhaps they also require a different evolutionary explanation.

Some readers might go further and argue that the arts are the *only* examples of true human uniqueness, as we the only species on the planet that practices them. While almost every other trait we have discussed in this book is uniquely human only in its quantitative excess, the arts appear qualitatively unique. This chapter seeks an evolutionary explanation for these remarkable phenomena.

Beyond Memetics

In the previous chapter, we defined culture as a dual process driven by basic memetic mechanisms. The first cultural mode involves essential survival knowledge learned from parents, relatives, and friends as part of the within-group heritage. The second mode is a signaling overlay that involves learning elaborate, but often trivial, social displays, which provide the basis for ultra-sociality. Both of these cultural modes are consistent with Dawkins's (1976) meme theory. Recall that memes represent functional connections, residing in brains, that link perceptions with imitative actions. Thus, a young hominid might watch a parent use a tool to harvest food, and then imitate the behavior to comparable effect. The memes associated with using tools, an essential survival skill in early human populations, thus propagate from parent to child, and down through the generations. Similarly, in encounters between different SKIF groups, traditionally a source of competitive hostility, enculturated humans engage in signaling displays that tend to moderate threats and reduce provocations. Each group may observe and imitate selected elements of the others' displays, in accordance with meme theory.

The question in the present chapter is whether the arts fit this meme-based cultural duality. At a superficial level, the answer appears to be yes. For example, a child might watch a master artist at work and then attempt to imitate the same basic techniques to produce new works. Similarly, the products of such artistic activities represent potential cultural signals that outside groups might observe and assimilate into their cultural traditions, thus promoting social unification between groups. Thus, at first glance, the arts appear consistent with basic meme theory and our dual model of cultural evolution.

Looking deeper into the problem, however, we come to a seeming contradiction. Specifically, if memes propagate through imitation and repetition, what happens when a talented individual performs an action so far beyond normal capabilities that almost no one else can imitate it? For example, the great Italian tenor Luciano Pavarotti had a voice that almost no one in the world could match. When he sang an aria, it might have inspired others to imitate, but the result usually fell far short of the original. Certainly, anyone hearing me attempt the same aria would hardly feel compelled to imitate it. According to meme theory, failure to imitate represents the extinction of the meme. Thus, in ancient times, how could the artistic talent of an ancestral (probably thinner) Pavarotti propagate through the generations if almost no one had the ability to imitate it? How does such rare talent endure when its practitioners represent a vanishingly small proportion of the population? Does simple meme theory account for this, or is there more to the story?

To find an answer, we need to understand the arts as an outcome of human evolution: biological, social, and cultural. This requires examining two basic issues: First, what causes the emergence of artists in a human population? Second, what drives the acceptance and appreciation of art and artists by the much larger population of nonartists?

The Emergence of Artists

The simplest explanation for the production of artists in a population is not memetic but genetic. Art is a product of the human mind. The mind is ultimately a function of the human brain, and the brain in turn is a product of the human genome, according to the materialist view adopted in this work. While cultural experience certainly adds much to the developing mind, genetics determines the neural framework in which the mind must dwell. Because genes vary across human populations, as they must in any evolving species, each of us starts life with slightly different behavioral and perceptual predispositions, on which the experiences of life build and elaborate.

In this view, true artists simply represent natural variants whose genetic gifts predispose them to extract, process, and express information differently than the rest of the population. For example, true artists may possess sensory augmentations that exceed the normal range of human abilities. They experience the perceptual world differently from everyone else. They see more (or differently), they hear more (or differently), and so on. If such variations occur naturally in human populations, as a result of genetic variations and their phenotypic expressions, then our species must always produce potential artists at some characteristic rate.

Artistic variants can take many forms. Because the human sensory cortex has disproportionately larger representations in the visual and auditory senses, random variations are perhaps more likely to produce visual and auditory artistry in our species. However, it is also possible for augmentations to occur in other senses: kinesthetic (dance), olfactory (perfumery), gustatory (cooking), and so on.

In some cases, these sensory variations may go so far beyond the normal range that the affected individuals actually have difficulty coping with the normal world. Indeed some psychologists attribute certain psychopathologies, such as schizophrenia, to the excessive augmentation of perceptual functions. The victims, unable to filter or process incoming information properly, are overwhelmed by their unique sensations and perceptions, leaving them unable to behave adaptively in normal social situations.

But perceptual augmentations represent only part of the artistic process. The expression of artistic talent also requires some matching ability on the action side of the nervous system. Budding artists may have unique sensory representations of the world, but they are not painters or musicians or dancers until they master the requisite motor skills with dexterity, coordination, and grace. According to the genetic model, individuals may vary as much in their potential motor skills as in their sensory processing. Some painters may have extraordinarily fine motor control of their hands and wrists. Some dancers may have exceptional flexibility and fast-twitch musculature. Such augmentations on the action side also represent a genetic predisposition toward artistry.

By this reasoning, the most extraordinary artists are those who have augmentations in both the perceptual and action domains. Such individuals not only experience the world differently, they have the requisite motor apparatus to act upon their unique perceptions to produce genuine works of art. Of course, by simple probability, it is more likely that individuals will have one or the other of these talents, but not both. Thus, more individuals may have exceptional gifts on the sensory side, but lack the physical ability to express their talent. Others may have exceptional physical tools as performers, but lack the perceptual creativity to produce new works of art. Thus, musicians and composers are not routinely interchangeable, nor are actors and directors, dancers and choreographers, and so on. Yet all may contribute to the arts in their particular areas of strength.

Assuming the requisite genetic enhancements occur with reasonable frequencies, there will always be a small population of potential artists in every human society. As these special individuals discover their talents and begin to express them, they may attract the attention of the wider community, beginning with their families. At this point, young artists may experience either support or condemnation for their unique abilities. If they are born into families of artists, as the genetic model suggests should be more likely, their early childhood environment will be more likely to nurture and develop their talent. In families of merchants or warriors, on the other hand, it may not. Thus, while the genetic model suggests that latent artistic talent may be present in certain individuals, the actual expression of such talent depends in part on the circumstances in which they live. Arguably, many more individuals have the requisite genetic foundations of artistic talent than there are recognized artists in the population.

As they continue their pursuits, artists must eventually find a place in the community outside their immediate families. Here, the human social structure plays a key role. Depending on the nature of the social group, artists may be warmly received, and therefore encouraged to flourish, or coldly shunned, and therefore forced to fail. How societies treat artists depends on the degree to which the members of the group appreciate art and support the artists that produce it. The next section addresses this crucial issue.

The Appreciation of Art

In most human societies, the appreciation of art and artists is a normal occurrence, something we take for granted. However, in evolutionary terms, the tolerance of artists is more problematic. Individuals who perceive the world differently than everyone else, and who act upon those perceptions differently, may represent a liability to a hard-pressed group. Consider, for example, a band of hunter-gatherers trying to eke out an existence in a hostile world of predators and food shortages. For such groups, art seems relatively useless. It does not produce food or water, build shelters, or fend off marauding predators and hostile raiders. The artist's inability to share the group's normal perceptions or to participate in their cooperative actions may doom him or her to a short, unproductive life. Natural selection, according to the theory of *Homo dominus*, should always favor the hominid pragmatist focused on control, rather than the artist focused on novelty or imagination. In the challenging world of the early hominids, what then was the value of art? How did those groups of rugged survivalists come to tolerate the presence of seemingly unproductive artists in their midst?

Again, the notion of genetic predispositions, this time in the brains of nonartists, provides a starting point. Several theorists (Aitken 1998; Coss 1968; Dissanayake 1992; Zeki 1999; Dutton 2009) have suggested that true works of art represent special patterns of stimuli that tap into certain powerful sensitivities in the brain of the observer. In other words, among all the possible patterns of stimulation present in the environment, a special few have the potential to trigger unique reactions in the observer. Such sensitivities may emerge genetically, as part of the observer's brain circuitry, or developmentally, as a result of their childhood experience. True works of art, according to art theorists, represent one particular class of such patterns.

To see this process at work, one need only watch infants and small children react to pictures, music, or dancing. We do not need to teach children what is interesting, funny, or moving. They react instinctively to it, as if predisposed to respond to such unique and novel sensory patterns. Indeed, for infants and children, almost anything new and different will produce such a reaction, as if, for the child, the world is utterly full of art. The adult artistic experience may be a reflection, perhaps even a developmental carry-over, of this same phenomenon.

Over time, however, as adult cognitive expectations become better established, our childhood wonderment tends to fall away. The art in everyday things becomes less remarkable, and it no longer provokes the same childlike reactions. The maturing mind seems intent on transforming novelty into expectation. The adult cognitive engine increasingly filters incoming information so that only the useful bits get through. Our wonderment and delight in the ubiquitous art of the world gradually gives way to a calculated control of perception. The bubbling dissonance that once dominated our internal world gives way to the cooler control repertoire necessary to survive in a hostile world.

Sometimes, however, our adult control filters fail to catch certain stimuli, and they trigger a moment of surprise, shock, or delight. That, I believe, is the essence of the adult artistic experience. True art seems to lie in the surprising mismatches between our perceptions and our expectations, which trigger bursts of childlike dissonance. From this perspective, we can find true art in certain surprising disruptions of our established control repertoires.

Because each of us has a unique set of perceptual capabilities and a unique deck of internal expectations, each of us has a unique definition of true art. For some people, such experiences may come visually in response to painting or dance, in others, audibly in response to music or stories, and so on. Patterns that do not trigger such cognitive mismatches, no matter how beautifully crafted, do not rise to the level of true art, as judged by one's own personal definition. Thus, what I see as art, you may consider trivial junk, and *vice versa*. The appreciation of art is less a function of finely crafted work than a personal experience. It happens whenever a sensation slips past our control filters and tweaks our adult expectations with a moment of childlike surprise.

Thus, the appreciation of art has an instinctive foundation. It is not something that needs to be taught. It is simply a part of our genetic heritage, augmented by cultural experience. This may help to explain why nearly all human societies, including those whose precarious existence depends on a rugged survival mentality, still accept and encourage artists and their works. The appreciation of art is thus an unexpected outcome of a controlling mind.

Mechanistic Interpretations

From the control perspective, then, the human artistic sense reflects the dissonance that arises when certain novel perceptions create mismatches with certain established reference expectations in the individual's brain. Such dissonance may be disturbing or delightful, sometimes even both. As we discussed earlier in Chapter 3, emotional reactions reflect the patterns of dissonance swirling in the observer's mind as they attempt to cope with their mismatched perceptions and expectations. Works of art certainly have such emotional effects.

Figures 9.1 and 9.2 demonstrate a mechanistic interpretation of these processes, including both the production of artists and the appreciation of art. Figure 9.1 illustrates the production of artists. This schematic depicts the same basic control module we have seen in previous chapters, particularly Figure 8.3 which showed the mechanistic interpretation of cultural signaling. In Figure 9.1, we take this same basic concept to the next level. The figure depicts the emergence of an expectation that is *entirely imaginary*. It does not exist in the real world, at least not yet. It is a perception that has never been perceived.

In other words, as the brain of *Homo aestheticus* evolved, the incremental increases in cortical mass began to form expectations that had no physical representation in the natural world. These emerging hominids began to live their lives in a world of imagination.

By the cold logic of control systems, however, expectations that have no perceptual match in the real world will produce a state of perpetual dissonance. Because normal actions cannot resolve such dissonance, the individuals who carry these novel expectations must explore actions outside their normal behavioral repertoires. This, I believe, is the essential driver of artistic activity.

Consider an example. A group of *aestheticus* hominids waits out the cold winter in a network of caves, such as those at Chauvet. As they sit through the long dreary months, some of them develop vivid dreams (expectations) of the animals roaming the plains and the forests just outside the cave entrance. However, their actual perceptions are limited to the grey cave walls, illuminated only by firelight and devoid of life. Most of the group simply has to endure this dissonance, as they have no practical way to resolve it. However, because this is a group of *aestheticus* hominids, there are true artists among them. These individuals have stronger expectations, which, in the cave's impoverished perceptual milieu, produce stronger dissonance, which in turn drives these artists to explore ever more elaborate and creative actions to moderate their dissonance. At some point, their actions slip out of the mundane and into the artistic. Some of them begin to act out scenes involving the animals familiar to them. Some vocalize or chant to recreate the sounds of their natural environment: animal calls, the wind in the trees, water flowing over rocks. Others pick up pieces of charcoal and deftly scrape the cave walls to bring the images of familiar animals into view. Through all these artistic modes, the familiar animals of the spring and summer suddenly take form within the confines of the cold stone walls. The more talented the artists are in their chosen media, the more real their subjects appear. In each case, however, the artist's creation attempts to match an expectation already present in the artist's mind.

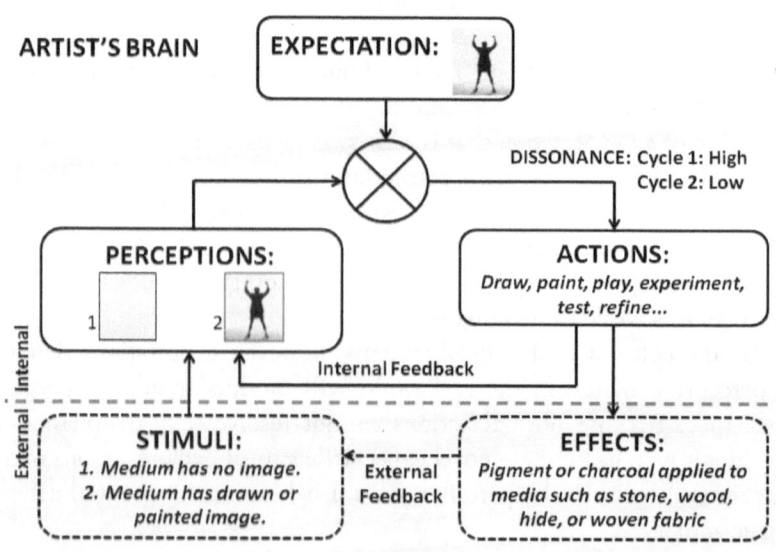

Figure 9.1. A mechanistic interpretation for the production of artists in a hominid society. The evolutionary expansion of cortical control architectures eventually produces novel expectations for which no external perception yet exists. This creates dissonance that can only be resolved by novel actions in the environment. These *artistic* actions are successful only when they transform an external medium to match the artist's internal expectation. The resulting object or display may be a work of art, depending on its effects on other observers (see Figure 9.2).

From a mechanistic perspective, the events in the cave are simply natural outcomes of a neural architecture that functions as a control system. Whether sequestered in a cave, or roaming a modern city, individuals seized by persistent dissonance between what is present in their minds, but absent in their world, will always tend to act in novel ways to create novel perceptions to match those internal expectations. For some, this becomes an artistic activity: a perceptual transformation that emerges from experiment, improvisation, and simple play with various elements of the environment. The outcomes are new patterns of light, sound, smell, taste, touch, and movement that progressively converge on the artist's internal expectations, and from which they achieve a measure of dissonance reduction. Before a work of art ever appears in the outside world, it exists as an expectation in the mind of the artist.

Once the work of art emerges in the outside world, however, it becomes accessible to everyone else in the community, revealing in tangible form the artist's once-private expectation. What happens next is the second fundamental element of artistic evolution: the acceptance and appreciation of art by nonartists.

Figure 9.2 provides a mechanistic view of what happens when nonartists encounter works of art. In effect, this is the reverse of what happens in the creation of art. Where the artist has a novel internal expectation, but no external perception to match it, the observer perceiving a work of art for the first time experiences a novel perception for which they lack any matching internal expectation. Thus, as the ancient cave painters transformed the cave walls, the rest of the tribe experienced the perception of familiar animals and scenes, but in an entirely novel form and place for which they had no established cognitive expectation. The resulting mismatch must have produced dissonance in their controlling minds, just as the reverse process did for the artist. And like the artists, this dissonance compels them to act in ways that resolve the mismatch.

Unlike the artists, however, who could actively transform their own perceptions by producing works of art, these first *art critics* had only two ways to resolve their dissonance: 1) by creating new reference expectations in their minds; or 2) by banishing the work from their perceptions. In other words, an individual encountering true art for the first time must either *stretch* their expectations to match those of the artist, or destroy the offending work of art, figuratively or literally. Both strategies resolve the mismatch, one by adding to the observer's deck of internal expectations, the other by shunning or destroying external perceptions.

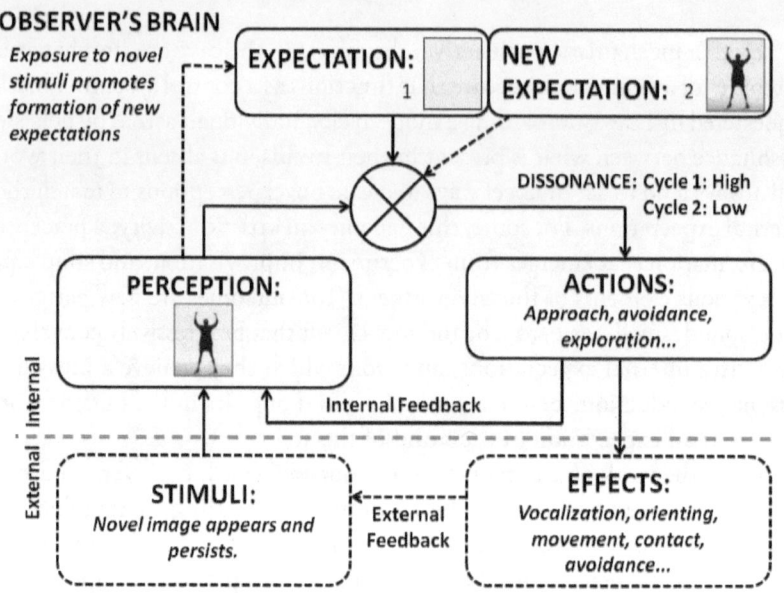

Figure 9.2. The artistic mechanism reverses to explain the appreciation of art by other observers (compare Figure 9.1). When an observer encounters a novel pattern or display, for which no prior expectation exists, there is a cognitive mismatch. This produces dissonance that persists until the observer's cognitive structure assimilates the novel pattern as a new learned expectation. True art therefore promotes the creation of new expectations in the observer's mind. Alternatively, the observer may reduce dissonance simply by avoiding or destroying the novel perceptual pattern.

In Chapter 8, we introduced the notion of expanding cognitive expectations in connection with cultural signaling. There we saw the adaptive advantages of utilizing the brain's expanding cortical mass to build a richer variety of alternative expectations for social displays. That same type of mechanism may help explain how exposure to true art can stimulate the production of new expectations. As individuals are exposed to novel stimuli and situations, they are challenged to assimilate new information and stimuli. This occurs primarily through a learning process that involves the construction, modification, and consolidation of memories encoded in neural connections. Depending on how the system encodes such memories, they may serve subsequently as expectations in the individual's control repertoire. Arguably, virtually everything we learn as adults has the potential to become part of our deck of expectations. As we saw in Chapter 1, such learned expectations are just as potent in driving human behavior as the instinctive ones. Thus, as new expectations emerge in response to a work of art, the work's original novelty moves gradually into familiarity. As the newly learned expectations take hold, they match the previously novel perception and null the observer's dissonance.

The fact that almost all human societies appreciate art in some form suggests that the strategy of expanding cognitive expectations is the path most often taken by the emerging human species. The alternative, ignoring or destroying art, has certainly had its practitioners in human societies at various times, but it hardly represents the general case. Far from shunning novelty, we humans appear to seek out new and different stimuli, constantly filling our minds with perceptions for which we have no prior experience and therefore no established reference expectations. That we seem to use such novelty as a way of expanding our internal reference decks suggests that the appreciation of art is a natural outcome of our cognitive control structure, as well as a natural source of stimulation to grow it. That a concept so simple as control could, in theory, propel a species into the heights of artistic achievement is truly remarkable.

Homo aestheticus and the Human Strategy

Characterizing the arts as outcomes of neural control processes is probably offensive to most true artists. I am, they might say, reducing the beauty of art and the passion of the artist to operational relationships on a schematic. In the process, I am taking something away from these extraordinary human activities. That is certainly not my intention. The advantage of using operational language to describe the creation of art and our response to it is that such language better accords with the evolutionary foundations of the human brain as a biological agent of information processing. That is, after all, what this book is about. The theory of *Homo dominus* argues that human beings perceive with the intent to control, so the arts must somehow fit into this model. In that context, we can say that the arts derive their power from their ability to surprise us, to playfully bump up against our established expectations and push us off balance, forcing us to try to regain control. The test of true art is its *stealthy* avoidance of our perceptual defenses, its mysterious ability to slip past our control filters to reveal something unexpected. As it penetrates our minds and emotions, art playfully mocks our sense of control, like a parent playing with a child. Had the human species not evolved such powerful control systems, we would not take such delight, or gasp with such shock, when true art slips unfiltered into our minds.

The evolution of *Homo aestheticus*, a name first used by Dissanayake (1992), ushered in both the creation of art and its appreciation by the community. From our ancestral gene pool, a few talented individuals began to emerge in every generation. Their brains were different, and the minds that resided in those brains carried different concepts of the world and its relationships. Their novel expectations forced these individuals to act in new ways act to transform their own perceptions. With skill and practice, plus a strong dose of courage and persistence, they invented the unique phenomena we call the arts. In time, the human world acquired a collection of objects, techniques, and displays that interested, intrigued, delighted, and shocked their owners and observers. That process continues today. The creation of art reflects the rare amplification of the human capacity for perception and action. The appreciation of art reflects the compelling value that artists bring as agents for expanding our understanding of what the world is and, more importantly, of what it could be.

CHAPTER 10

Homo mortalis, the Mortal:
Controlling the Uncontrollable

Stepping Between, Never On, the Lines

After a short career in the Detroit Tigers organization in the 1960s, George Gmelch gave up professional baseball and took up cultural anthropology. Certainly not the most common career transition, but fortunate for us as it created a unique bridge between two seemingly disparate walks of life. Gmelch's experience and inside knowledge of professional baseball provides a rich source of material for his observations and insights as social scientist. Particularly interesting in the present context are his observations of the bizarre and often comical superstitions that seem to inflict baseball more than other sports (Gmelch 1978, 2001). Indeed, after fractured syntax (Berra 1998), weird superstitions may be baseball's most enduring contribution to society. Some of Gmelch's examples:

> On each pitching day for the first three months of a winning season, Dennis Grossini, a pitcher on a Detroit Tiger farm team, arose from bed at exactly 10:00 A.M. At 1:00 P.M. he went to the nearest restaurant for two glasses of iced tea and a tuna fish sandwich. Although the afternoon was free, he changed into the sweatshirt and supporter he wore during his last winning game, and one hour before the game, he chewed a wad of Beech-Nut chewing tobacco. After each pitch during the game, he touched his letters (the team name on his uniform) and straightened his cap after each ball. Before the start of each inning, he replaced the pitcher's rosin bag next to the spot where it was the inning before. And after every inning in which he gave up a run, he washed his hands.

When asked which part of the ritual was most important, he responded, "You can't really tell what's most important so it all becomes important. I'd be afraid to change anything. As long as I'm winning, I do everything the same. Even when I can't wash my hands (this would occur when he had to bat), it scares me going back to the mound. I don't feel quite right."

Other examples:

New York Yankees Wade Boggs eats chicken before every game (that's 162 meals of chicken per year), and he has been doing that for nine years. Chicago White Sox pitcher Jason Bere listens to the same song on his Walkman on the days he is to pitch. His teammate, Ozzie Guillen, doesn't wash his underclothes after a good game. San Francisco Giant pitcher Ron Bryant added a new stick of bubble gum to the collection in his bulging back pocket after each game he won. Jim Ohms, my teammate and pitcher on the Daytona Beach Islanders, put another penny in the pouch of his supporter after each win. Clanging against the hard plastic genital cup, the pennies made an audible sound as he ran the bases toward the end of a winning season.

Although his anecdotes are humorous, Gmelch's work serves a serious scientific purpose. Baseball superstitions mirror human behavior displayed almost universally across the world's cultures and, most paleoanthropologists believe, historically through the generations. Gmelch uses the general term *magic* to encompass this collection of superstitions, rituals, taboos, fetishes, and other oddities. Such behavior is important in anthropology, according to Malinowski (1948) and others (Tylor 1871; McLennan 1876; Frazer 1922), as it reflects the emergence of spirituality in the human species and provides a steppingstone to the cultural development of religion. This chapter deals with the emergence of these forces and their roles in human life. While baseball superstitions seem almost religious in their hold on certain players, and the American film *Bull Durham* even refers to the "church of baseball," our brief trip to the ballpark will eventually lead us into deeper matters of life and death.

Uncertainty

Arguably, the whole point of a control strategy is to reduce uncertainty. Human behavior reflects a relentless drive to eliminate uncertainty and fear and to substitute predictability and confidence. Other species do not do this to the degree humans do. They simply accept uncertainty, or more precisely, they remain blissfully unaware of alternatives to an uncertain existence.

Uncertainty and control are essentially polar opposites. When control increases, uncertainty decreases. Where uncertainty remains intractable, control is impossible. This latter condition is important in the present analysis. What does a controlling species do when confronted by uncontrollable events? How do we reduce uncertainty when uncertainty is the only certainty? Such questions are crucial to understanding the human evolutionary strategy, for we need to know what the human species does when that strategy fails. Fortunately, cultural anthropologists have been looking for answers for many years. Their explanations, like those of many professional baseball players, center on a remarkable set of behaviors, which I will refer to generally as *superstitions*. They represent the outcomes of humans grappling with intractable uncertainty. In the following pages, I shall try to connect them to the theory of *Homo dominus*.

In his baseball observations, Gmelch noted that superstitious behavior occurred more often in situations where the element of chance was greater. Baseball has three main activities: hitting, pitching, and fielding. (We will ignore spitting and scratching for the moment.) In two of these, hitting and pitching, even the best execution often has a negative outcome. A hitter can crush the ball, only to have it fly directly into the glove of a strategically placed fielder. Similarly, a pitcher can throw the ball to a precise location with the desired the velocity and movement, yet still see the batter bloop the ball over the infield for a game winning hit. Fielding, on the other hand, is more predictable, at least at the major league level. Professional players usually convert about 99 percent of their chances into actual outs. If a fielder can reach a batted or thrown ball, it usually yields some positive result, either getting an out or limiting the other team's scoring opportunities. Gmelch showed that strong superstitions tended to surround hitting and pitching, but were less prevalent in fielding.

This pattern is remarkably similar to that described by Malinowski (1948) in an entirely different setting. He observed fishing practices by the natives of the Trobriand Islands off New Guinea. The islanders have a choice of fishing the inner lagoons, where the waters are relatively safe and the catches are small but predictable, or venturing out on the open sea to go after bigger and more abundant fish. The latter, however, exposes them to greater risks from sudden storms, rough seas, and marine predators. Moreover, the catch is less predictable, sometimes yielding full nets, other times nothing. Malinowski noted that these two kinds of expedition involve very different preparation rituals. In general, the islanders rarely resort to their *fishing magic* when plying the safer waters of the inner lagoons, but regularly invoke such rituals when venturing out into the open ocean. Malinowski concluded:

> We find magic wherever the elements of chance and accident, and the emotional play between hope and fear, have a wide and extensive range. We do not find magic wherever the pursuit is certain, reliable, and well under the control of rational methods.

Thus, the Trobriand islanders, like American baseball players, tend to resort to superstitious behavior whenever they engage in activities with high uncertainty. This is particularly true, as Malinowski notes, when the situation arises repeatedly and has important consequences. Repetition is important because it confirms the uncertainty of the situation. Consequences are important because they define the seriousness of the situation. In professional baseball, the game represents the players' livelihood. Failure to execute could mean financial losses and reductions in status. For the Trobrianders, the consequences of failure in fishing the open sea could mean hunger for the village and perhaps death for the fishermen caught in a storm or attacked by sharks. Thus, in an enterprise with inherent uncertainty, repeated regularly and having significant consequences, human beings engage in superstitious behavior. Anthropologists have documented such patterns almost universally across human cultures.

Death

There is another class of events that promotes superstitious behavior, but for which uncertainty has a quite different meaning. Indeed, the outcome of one such event is entirely certain: death. Death poses a major problem for a controlling species, not just for the obvious reason, the cessation of life, but because it demonstrates the ultimate futility of the control strategy. Death is the quintessential loss of control. When this singular event overtakes us, we lose all power to maintain the carefully constructed edifice of our life. We can no longer fulfill our daily expectations, protect our loved ones and provide for their needs, commune with our friends, and so on. All of these fall away when death renders us immobile and unresponsive to the outside world.

Humans have an awareness of death that other species do not have. Our brains are better equipped to observe, remember, and adapt to repeating patterns throughout the course of our lives. As social animals, we learn from each other, in part by observation and imitation, in part by linguistic communication. From observations of the death of others, our cognitive architectures can project to future situations. From stories, we can conjure the specter of death without actually experiencing it. These experiences bring death into our cognitive calculations as an enduring and ultimately inescapable threat. No other species is cursed with this premonition.

The foreknowledge of death creates intractable dissonance in the controlling human mind. This comes from our instinctive expectation of immortality mismatched with a growing perception of mortality. This fundamental mismatch creates unresolvable dissonance which no rational action can ever null. Thus, our natural propensity for control inevitably falters when confronted by the absolute impossibility of controlling death. As we shall see, this terrible contradiction is the primary impetus for a rich repertoire of uniquely human superstitious behavior.

In youth, humans solve the problem of death by denying it. Young people assume a sense of immortality that manifests itself in the denial of death as a relevant event. They proceed actively and aggressively, often oblivious to the potential danger that adults, particularly parents, perceive at every turn. In the young mind, death exists only in the abstract, a theoretical construct that does not command much practical attention. Children learn the safe way of doing things more to please their parents and teachers than from any clear image of fatal consequences. In these early stages of cognitive life, death represents a curiosity rather than a projected certainty.

Over time, however, death develops a relevance that cannot be ignored. Everyone remembers his or her first experience with death, often a pet or another unfortunate animal killed accidentally or deliberately in a hunt. We remember the death of a relative and its impact on our parents. We remember our first experience with the death of a peer, someone we knew or grew up with. With each passing year, the toll grows until we begin to understand that death is an event of increasing likelihood, applicable to ourselves as well as others. This realization becomes particularly strong as we move into the reproductive phase of our lives when we take responsibility for raising children.

The inevitability of death is true for all species, of course, but it is something only the human mind appears to experience beforehand. Our nearest genetic relatives, the chimpanzees, certainly have the same exposure to death. Chimpanzees age and die. Members of the social group may succumb to disease or accident. These events appear to trigger grief and depression in chimpanzees (Goodall 1986). However, primate researchers find little evidence to suggest that chimpanzees comprehend the inevitability of death or take any action if they do (Leakey and Lewin 1992). Like most of their behavioral repertoire, the chimpanzee's response to death appears reactive, as if it always comes as a bewildering surprise. If they have intimations of their own mortality, their overt behavior does not seem to reflect it.

In humans, as death becomes a cognitive reality, our controlling instinct is to manage the perceptions that surround it. On the one hand, there are practical matters. For example, death produces a corpse. As a social species, humans have expectations of how other humans should behave. A corpse violates these rules. It looks like a human being, perhaps someone we knew and loved, but it no longer behaves like a human being. It does not move or breathe or react. It grows steadily colder to the touch and the joints stiffen. For a time, we can deny its reality and treat it as merely as a person sleeping, our closest approximation to normal behavior. Eventually, however, our dissonance must increase as the *sleep* continues well beyond normal expectations. In time, the situation becomes acute as decomposition sets in. The cessation of metabolism leaves the human body defenseless against scavenging by bacteria, insects, and other parasites. They feed unchecked, producing physical disfigurement and the distinctive odors of decay. Most humans show aversive reactions to such odors, including physical nausea, vomiting, and avoidance (Rozin, Haidt, and McCauley 1993). Biologists believe these reactions represent an instinctive defense against the dangers of decaying animal flesh. Thus, the threat posed by the human corpse compels the survivors to take defensive action. Somehow, they must make the corpse disappear. Boyer (2001) considers the necessity of dealing with decomposition as a fundamental transforming experience among the survivors.

Paleontologists generally believe that funeral practices among the early hominids were probably rudimentary, perhaps as simple as abandoning the corpse and moving on. However, as the human species evolved, particularly with the advent of cultural learning, the processes involved in handling corpses became more elaborate. The initial knowledge underlying funeral processes probably came from observations of death in the natural world. For example, as predators and scavengers, humans would have had considerable experience dealing with the carcasses of animals. They would also have witnessed the handling of dead animals by other predators, for example, the behavior of big cats that bury carcasses or deposit them in trees. Indeed, some hominid observers might have seen members of their own group subjected to such treatment. Thus, the practical models of burial, elevation, or isolation of corpses may have been familiar even to the earliest hominids.

We do not know precisely when intentional funeral practices began among the hominids. The major archeological evidence dates back about 100,000 years ago for Neanderthal burials (Tattersall 1998, 1995; Trinkaus 1983), but recent discoveries in Spain suggest intentional burials among *Homo heidelbergensis* that date back perhaps 350,000 years ago (Bischoff et al. 2003). Neanderthal gravesites show clear evidence of ritual interment, including the east-west orientation of the cadaver, deliberate arrangement in a sleeping or fetal position, and the placement of flowers, crafted objects, or animal bones in the grave. In some cases, the presence of pigments such as red ochre suggests the ritual decoration of the body. The Neanderthals qualify as an advanced human species that predates modern humans by about 250,000 years. Their cultural sophistication is under considerable debate, but clearly, their funeral practices reflect a certain cultural tradition with respect to death.

No clear evidence exists for ritual burials among more ancient hominids. There are some hominid fossils dating to 1.4 million years ago that bear apparently deliberate cut marks, as if the body underwent some postmortem processing (Pickering, White, and Toth 2000). Whether this had any ritual or funeral significance remains uncertain.

In addition to the physical challenges of dealing with a corpse, death also presents certain psychological challenges. As we have seen in previous chapters, human social systems incorporate a suite of powerful attractive forces involving pair-bonds, parent-child relationships, and a wide variety of kinship/friendship links. Such relationships involve a rich array of expectations, some developed over many years. These expectations become relatively permanent fixtures in the cognitive structures of the social group. With the death of an individual, however, these longstanding expectations no longer have any possibility of finding a perceptual match. The survivors, therefore, must experience strong dissonance that they cannot resolve by any familiar action. This dissonance may propagate across the entire social group, affecting even those who were not acquainted with the deceased. The net effects can be serious, sometimes causing significant mental and physical deterioration, erratic behavior, and a general disruption of the social system.

The specter of death clearly represents an acute challenge for the human species, both physically and psychologically. As we have seen previously, human cognition builds on the instinctive belief that action can always alter perceptions to resolve dissonance. The realization of death's inevitability severely challenges that ancient conviction.

Spirituality

So what does the human mind do when confronted by a problem like death? A story about the great comedian W. C. Fields provides an answer. When Fields was gravely ill and not expected to recover, one of his friends visited him in the hospital and found him reading the Bible. Surprised at such uncharacteristic religious behavior, he asked Fields what he was doing. Fields replied in his classic style, "Looking for loopholes." This line is funny and poignant, but it also says something basic about how humans deal with the problem of death, whether imminent, as in Fields' case, or more distant, as we all hope for ourselves. Whenever we encounter problems that we cannot resolve by our basic control approach, we look for loopholes. Our minds simply cannot accept the failure of control, so we resort to inventing perceptual surrogates that allow us to regain a sense of control, whether real or imagined. Thus, when the outcome is seemingly certain but uncontrollable, like death, we instinctively search for areas of uncertainty, in the hope that these may provide loopholes through which we might regain control. Thus, this "loophole tactic," if we can call it that, involves first making the certain uncertain, and then dealing with that uncertainty to make it certain again, but on our own terms. Such convoluted logic is a uniquely human cognitive invention, but it reflects, I believe, a deep-seated genetic predisposition. The term I shall use for this predisposition is *spirituality*, from which much interesting and problematic human behavior flows.

Human spirituality works like this. While death is certain, its consequences are not. In truth, they are unknown, because none of us has actually experienced death and its aftermath. It is therefore an area of uncertainty. In the controlling human mind, this presents a perceptual loophole: if we cannot control the event itself, perhaps we can control its perceived consequences. Thus, our cognitive control circuits begin to inject imaginary elements into our perceptions about death. For example, we try to see death not as the end but as the beginning; we envision a new life that begins after death, and so on. As these imagined perceptions intercalate themselves into our actual perceptions of death, they stimulate the formation of alternative expectations about the possible consequences of death. Just as we saw in Chapters 8 and 9, the human cortex supports the proliferation of learned expectations, including those that have no actual correlates in the physical world. These newly synthesized expectations in turn promote new actions in the form of rituals and superstitions that provide additional cognitive feedback to help sustain the imaginary perceptual transformation. Thus, by the expedient of creating a rich array of new expectations in the controlling mind, death moves from an entirely certain event over which we have no control to a relatively uncertain event over which we imagine considerable control. This elaborate cognitive transformation restores our sense of control, albeit artificial, and thereby provides us some relief from the dissonance of pending oblivion.

I use the term *spirituality* to describe this general human predisposition to build hypotheses about the unknown. Although most people affiliate the term spirituality almost exclusively with religion, I believe it has applicability to all problems, religious and secular, that have no immediate resolution in fact or logic. Confronted by such intractable problems, our controlling minds respond instinctively with hypotheses. We synthesize imaginary forces and objects that solve the problem, and then animate them in some imaginary way to produce the desired perceptual outcomes. True spirituality makes no distinction between the real and the surreal because it works entirely within the realm of imagination. It allows us to invoke hypothetical entities and forces outside our normal experience, indeed beyond all known physical laws, and then to proceed as if we had control of them. This capacity for spirituality, I believe, represents a fundamental genetic predisposition embedded in the cognitive architecture of the human brain. Several recent works have explored connections between spirituality and neuroscience (Beauregard and O'Leary 2007; Sounds True 2008).

Figure 10.1 illustrates a mechanistic interpretation of this process. Here the familiar control module shows perception interacting with expectation to produce dissonance. Initially, the dissonance arises from the mismatch between our expectation of immortality and our perception of impending death. This dissonance cannot be sustainably resolved, because nothing can prevent death from occurring at some point and, in the normal human mind, nothing can alter the expectation of immortality. Intractable dissonance is the inevitable outcome of this mismatch.

Now, if the only option for resolving such dissonance were action in the physical realm, that is, to *actually* overcome death as a physical reality, the human control system would have no way out of its dissonance. However, Figure 10.1 shows a way out of this trap. It is the internal feedback loop or "imagination connection" as William Powers (1973) called it in his original description of perceptual control theory. In cognitive control systems, these internal loops provide immediate internal feedback from the action side of the system to the perceptual side. This feedback initiates immediate changes in the internal perceptual mix. In Chapter 2, I suggested the importance of this connection in prediction, that is, the projection of potential consequences of action in advance of actual consequences. In the present context, the signals that pass from the action side of the system back into the perceptual side create imaginary transformations of the perceptions arriving on the perceptual side of the structure. Such transformations are at the heart of human spirituality.

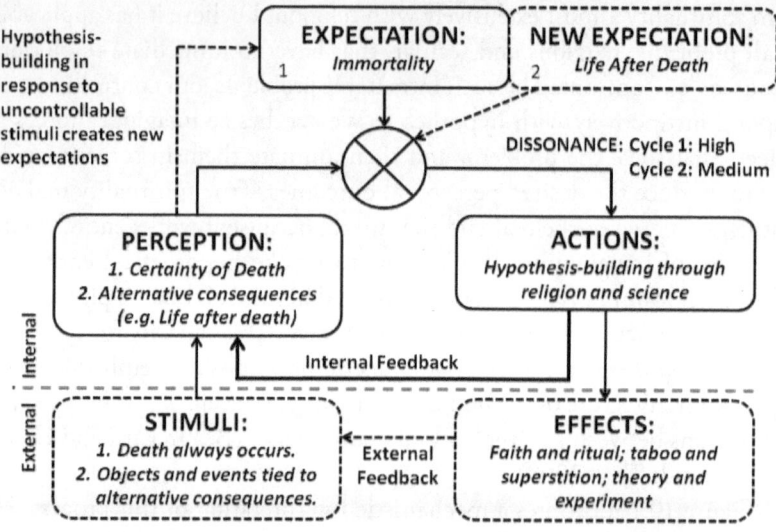

Figure 10.1. Control schematic showing the intractable dissonance that arises when uncontrollable perceptions conflict with fundamental expectations. Here the basic drive to survive conflicts with the perceived inevitability of death. This produces dissonance that ordinary behavior cannot resolve. The system responds by building *hypotheses*. These actions do not alter the actual occurrence of death. However, they may create internal cognitive feedback that alters the perception of death. These novel perceptions may contribute to the formation of alternative expectations, such as the concept of *life after death*. Hypothesis-building creates and sustains the illusion of control. The human predisposition to deal with unknowns and uncontrollables through hypothesis-building and superstitious behavior reflects the genetic predisposition of Spirituality, from which both religion and science derive.

In essence, the feedback loop depicted in Figure 10.1 functions as a kind of hypothesis-generator, which has the effect of producing a mix of new perceptual variations. As such, it stimulates the conversion of uncontrollable realities into more pliable hypotheses. The logic goes like this: As an individual suffers the mismatch between the reality of death and the expectation of immortality, intractable dissonance drives persistent but futile actions, both actual and virtual. Such actions, although ineffective in resolving dissonance directly, generate a constant barrage of *imaginative* feedback activity, which in turn produces various transformations of the perceptual mix. In the experiential realm, we might call these transformations dreams, visions, or hallucinations. However, even though they are virtual entities, they can have real perceptual effects, that is, their activity helps create new neural patterns laid down in memory as new expectations. For example, an active imagination may transform the expectation of death as *unconditional oblivion* into the less onerous expectation of *life after death*, involving the hypothetical event of post-mortem reawakening. As this spiritual feedback engine continues to spin out such hypotheses, it stimulates the proliferation of new learned expectations, again entirely imaginary, but sustainable as potential alternatives to the expectation of unconditional immortality. As these new expectations about death and its consequences become active, they may find a stable match with imagined perceptions, which may in turn reduce the dissonance associated with impending death. Once the system finds a path that reduces dissonance, it will tend to default to that path, since dissonance reduction is the fundamental operational imperative for such systems. Thus, even in a supposedly rational cognitive control system, like a human mind, a reality like death, which offers *no* hope of control, will always give way to hypotheses that offer *some* hope of control. As the system stabilizes, the newly synthesized expectation may become the default state, even though it originated purely as a hypothesis. Its moderating effect on dissonance promotes its acceptance as the preferred version of reality.

Of course, imagined perceptions and expectations generated by the human spiritual engine are not substitutes for actual transformations of death in the external world. They are mere approximations invented to make the dissonance of life under the specter of death more bearable. This means, of course, that there always remains an element of uncertainty about whether these hypotheses are actually true. And uncertainly stimulates and sustains superstitious behavior. Thus, in addition to the rich set of internal cognitive hypotheses that mask the unrelenting dissonance of uncontrollable problems, the human spiritual engine also promotes a rich set of superstitious behaviors that help sustain such hypotheses.

In the matter of death, then, the spiritual predisposition embedded in the human cognitive engine finds stability in hypotheses about an afterlife in which the dead retain control. Such hypotheses do not deny the event of death, which would clearly contradict reality, but instead offer alternative constructs about the consequences of death. To maintain the necessary internal feedback to sustain such hypotheses (i.e., to keep the loops closed), the bearer must continue to act in ways that promote the appropriate feedback patterns. Because the control system evaluates such actions entirely against *imagined* outcomes in the internal cognitive world, rather than tangible outcomes in the external physical world, they are clearly superstitious behaviors, as defined by Gmelch, Malinowski, and many others. The spiritual engines that generate such behaviors exist almost universally across human cultures (Boyer 2001).

Thus, with the emergence of *Homo mortalis*, the first spiritual hominid, human evolution created a mechanism for dealing with perplexing problems that offered no rational opportunity for control. The innovative human mind, so dedicated to controlling events, actually found a way to control the uncontrollable, at least in the domain of the imagination. The impacts of this spiritual engine on human cognitive, social, and cultural life were profound. Over evolutionary time, starting from the spiritual spark in the first *mortalis* hominids, there eventually grew two great cultural edifices: Religion and Science. Although we often characterize them as competitors, they are both products of the same spiritual predisposition. Both address the unknown. Both involve the creation of hypotheses. Both specify actions that promise control over important and seemingly uncontrollable events. Although spirituality is a fundamental biological property of the human mind, its overt expressions may take either religious or scientific form, as the following sections will describe.

Religion, Science, and Sibling Rivalry

Religion and science have always had an uneasy relationship. Even though both claim to do the same thing—to provide solutions to difficult human problems—they have always competed with one another. Like sibling rivals, both vie for our attention, and like petulant children, both are prone to tantrums when the other appears to get more attention. It has probably been this way since ancient times.

Religions tend to claim precedence over science in spiritual matters because religions are willing to tackle the truly difficult and frightening problems that confront us, like death and its consequences. Religion reminds us that people still die, and when they do, something happens to them. It is a frightening prospect for which human beings naturally seek some solace or explanation. Science is generally reluctant or unable to comment on such matters. But lacking rational scientific answers, the spiritual being will accept irrational ones, whenever they have the power to reduce dissonance. Thus, religion continues to thrive, and rightly so, given the true nature of humanness.

For its part, science argues that we cannot achieve real human progress by building on hypothesis and superstition alone. Although science actively promotes hypothesis-building as part of its repertoire, it insists that its hypotheses remain merely interim constructions to facilitate real solutions. Only by testing actions against actual perceived outcomes, as opposed to pure imagination, can we achieve actual control. Thus, even though science and religion share a propensity for hypothesis-building, the scientific mind does not stop with faith, but continues to test and evolve. Religion, science contends, tends to cling to untested beliefs and superstitions that cannot possibly convey the promised power of control.

While science may be guilty of needlessly denigrating the benefits of religion, religion also tends to go astray when it attempts to indoctrinate the faithful against the legitimate benefits and discoveries of science. Some religions consider knowledge a threat to faith. In 1633, for example, the Catholic Church threatened the torture and execution of Galileo Galilei, as they had done a generation earlier to Giordano Bruno, unless he recanted his sin of describing the orbits of Jupiter's moons (Reston 1994). More recently, the famous Scopes trial in 1925 pitted fundamentalist Christians against proponents of Darwin's theory of evolution. John Scopes, a high school teacher, joined a group of civil libertarians seeking to overturn a Tennessee statute forbidding the teaching of any view contrary to the biblical doctrine of creation (Larson 1997). After a sensational trial, the court found Scopes guilty and fined him a hundred dollars. A higher court later overturned the conviction, but did not overturn the law itself.

Interestingly, in the three hundred years between these two episodes, the Catholic Church itself underwent a transformation, its own evolution if you will, to the enlightened view that science does not necessarily demean faith, or reduce the role of religion in the lives of the faithful. The Catholic Church joined with many other faiths in the landmark *Voices for Evolution* (Matsumura 1995), an endorsement of peaceful coexistence between evolutionary theory and religion. Such developments represent strong cultural signals that religion and science both have valid roles to play in the lives of human beings. When either departs from its primary focus, human society is poorly served. It seems unlikely that science will ever deal effectively with the afterlife, and similarly, religion cannot help its adherents by thwarting the scientific exploration of the unknown or denying the products of human intellect. Science does not destroy faith and faith does not depend on ignorance.

In general, most human beings do not care much about the conflict between science and religion. They simply want answers that give them control over their lives. When science works better than religion, people embrace it. When religion works better than science, people embrace it. The pragmatic human mind seeks only control, which it will accept from any credible source.

The First Inklings

If spirituality reflects the controlling human mind confronting uncontrollable problems, then spiritual expressions must date well back in human history. As the human brain evolved, the early hominids would have experienced the frustration of wanting, but not having, control over large segments of their lives. Knowledge of death presented them with the prototypical spiritual problem, but it would have been only one of many uncontrollables confronting them. The early hominids lived at the mercy of powerful forces they did not understand. Torments such as droughts, floods, storms, wildfires, parasites, and predators plagued their lives and minds. Their rudimentary knowledge and limited technology offered some answers but not nearly enough. Faced with such ubiquitous problems, the ancient spiritual hominid, *Homo mortalis*, would have generated hypotheses about all of them. Each such hypothesis might have included supernatural forces (gods or spirits) that held sway over the particular elements of the problem: water, land, fire, air, life, death, male, female, and so on. This pantheon of ancient spirits, probably dating well back in hominid evolution, embodied the core hypotheses of the first animistic religions (Frazer 1922).

Once the hominids began hypothesizing the existence of gods and spirits as explanations for complex and uncontrollable things, they probably also began to use such gods and spirits as tools with which to gain some measure of personal control. After all, the spirits had the power to move the air and water, to create children and keep them healthy, to take people into the afterlife when they die, and so on. Such power would certainly be a source of envy to the would-be controlling species. Therefore, superstitious behavior among the early hominids probably dealt less with worshipping such spirits in the abstract and more with positioning themselves favorably to use the spirit's power to control events in their own lives. Controlling the spirits, understanding what they wanted, how they operated, what to do to curry favor with them, may have formed the basis of the earliest polytheistic religions.

As human social behavior evolved toward the new pattern of ultra-sociality, polytheistic religions also offered the flexibility and diversity needed to bring groups together. In the early stages of social intermingling, xenophobia remained a potentially divisive force, as outlined in Chapter 8. Polytheistic religions offered the opportunity for cultural signaling among wary tribes. As they intermingled, each arriving group would have contributed its own distinctive pantheon of gods and spirits to the new cultural mix. Other groups might have found similarities between these new spirits and their own familiar ones. For example, the god of fire might have had many names among the separate tribes, but all might have recognized some form of that particular god in their intermingled societies. Polytheism thus offered a permissive framework that accommodated the needs of diverse and otherwise competitive groups coming together for the first time. As groups merged, they built their religious systems from the polytheistic menu, ordering à la carte, so to speak.

Paralleling the evolution of religious culture, the early hominids also engaged diligently in scientific and technological development. The steady upward ratcheting of knowledge and technical expertise, as outlined in Chapter 5, gave the ancient hominids a growing scientific foundation for their daily lives. Rather than relying on superstition and ritual to curry favor with a particular spirit, the technological hominid, *Homo habilis*, developed rational processes for making things happen. Fire, for example, came from two wooden objects rubbed vigorously together or from the sparks generated by striking particular stones together. In the camps of *Homo habilis*, therefore, one no longer needed to pray to the fire spirit to send a bolt of lightning. As the hominid societies gained more technological control over various physical problems, they needed fewer gods. In time, some tribes shed most of their polytheistic spirits in favor of the more streamlined forms of modern monotheism.

While spirituality is a property of individuals alone, their willingness to share their personal hypotheses about the unknown, and then to act together to promote them, define the cultural process of religion building. Religions stand or fall primarily in how they handle the major unknowns facing human beings, of which death and its consequences surely sit at the top of the list. Spiritually, a religion succeeds by convincing the faithful that its hypotheses and associated actions bring control over the consequences of death. Culturally, a religion succeeds by proliferating its hypotheses and prescribed behaviors to a substantial portion of the population. While spirituality opens all human beings to such hypotheses, their proliferation depends on individual choice. Religions that offer hypotheses that are more believable, and prescribe behaviors that are more easily imitable, tend to succeed and grow, according to basic cultural memetics. Those that feature outlandish assertions or make extreme demands on the faithful (e.g., human sacrifice) typically find fewer adherents.

213

Most religious systems focus on teaching children a particular set of hypotheses, on the assumption that such beliefs will find almost permanent acceptance. Children live in a world of unknowns and uncontrollables, so arguably their openness to spiritual messages is considerably richer. Children also tend to accept beliefs from their parents and teachers often without questioning or testing until very much later in life, if at all. Children, however, have less experience with the inevitability of death, which gives them some immunity to the dissonance of mortality. Children often adopt the outward expressions of religion taught by their parents and peers, yet remain confident in their own immortality. Religion seldom takes its adult form until a person realizes death's inevitability. At that point, the individual may accept the previous teachings of parents and peers, or choose alternative hypotheses that better fit their adult needs.

The memetic nature of religions, as cultural phenomena, has prompted some critics, perhaps harshly, to invoke the notions of viral memes as an explanation for the spread of religion, specifically the aggressive evangelical varieties (Dawkins 1993; Brodie 1995; Lynch 1996). In some doctrines, the quality of the promised afterlife depends on the vigor of one's evangelical work. Those who fail to bring in new converts find themselves in danger of losing their preferred position in the afterlife. Such *quid pro quo* arrangements suggest viral memes. The doctrine's promise is tied directly to its own propagation. Such viral religions prey upon natural human spirituality and take advantage of an innate need to gain control over seemingly uncontrollable events. However, as with other viral memes, the religious *quid pro quo* survives only insofar as it balances effort with reward. The perceived benefit of the doctrine must balance the effort of propagating it, or the individual will simply drift to a less demanding religion.

The occasional misuse of religion for power and political purposes does not negate its considerable social benefit. In most societies, religion represents a stabilizing force that promotes morality and altruism, two cultural themes that help counter selfishness and unchecked aggression. By codifying unselfish behavior in a cultural form, religion facilitates the intermingling of SKIF communities into a harmonious cultural mix. By changing the human focus from territoriality and aggression to concerns about death and the afterlife, religion promotes shared values across competitive groups. Ideally, religion should deal exclusively with such matters, promoting our shared experiences as a species. However, in the flawed world of human beings, religious institutions sometimes drift toward controlling the faithful rather than simply serving their spiritual needs.

Homo mortalis and the Human Strategy

The penalty for developing a brain with the power to see the future is the vision it brings of our own mortality. For a species built on the need to control life's events and their consequences, death is a cognitive problem of the first magnitude. *Homo mortalis* was the first to understand this. This hypothetical species displayed the first inklings of human spirituality embedded in the cognitive control structure of their brains. When confronted with seemingly uncontrollable forces and events, these hominids began to elaborate hypotheses about what might be true, then to behave as if those imagined truths were fact. In time, they invented both science and religion to deal with their dissonance, using both to convey a sense of control, whether real or imagined. Even though both stem from the same spiritual instinct, religion and science have traditionally competed with one another, probably from the earliest human time. We can easily imagine ancient societies in which the shaman priests argued incessantly with the hunters and toolmakers over the legitimate boundaries of theology and technology. The general population simply yearned for clarity.

With the foreknowledge of death, *Homo mortalis* might have tumbled onto the path of anarchy and chaos, perhaps lapsing into the selfish aggression of *Homo bellicosus*. Knowing our eventual fate, why should anyone adopt a moral code? Why not simply indulge one's selfish impulses in the little time we have? The answer appears to lie in the simple evolutionary truth that if we did not follow a spiritual path, death would claim us sooner rather than later, perhaps without having successfully reproduced. From the biological perspective, death without issue truly means oblivion, whether foreseen or not. The abandonment of moral order in the face of certain death therefore represents a loss of control not a gain, something the human species always resists.

Expressions of spirituality pervade human society. We do not know precisely when they began, but if the control hypothesis is correct, then the spark of spirituality must have ignited at the moment our ancestors committed themselves to controlling their world, rather than simply enduring it. As we gather evidence for the human strategy, we must ask how the earliest hominids would have dealt with problems they could not control. How did they view the forces of nature? How did they deal with death? The answer lies in the simple hypothesis that when they started trying to control their lives, they surely started trying to control their deaths.

CHAPTER 11
Homo sapiens, the Wise: *Quo Vadis?*

The Path to Homo sapiens

Earlier in this work, I challenged the traditional name of the human species, *Homo sapiens*. Far from behaving with wisdom, justice, and sensibility, our species survives on a diet of control. We move and shape the world to conform to the set of expectations residing in our brains. This survival strategy has sustained our species from its humble beginnings on the margins of the ancient rain forest, through a series of impressive migrations, into the present age. From our hardscrabble existence as the forgotten runts of the forest, the hominids have now risen to threaten almost every life form on the planet with extinction, either deliberately, through attack and exploitation, or inadvertently, through habitat destruction and pollution. Only the most prolific and diversified species—viruses, bacteria, flatworms, insects, and so on—seem able to stand against us. In time, perhaps they will overwhelm us and turn our vision of control into a ghostly illusion. For the moment, however, the human species, *Homo dominus*, holds a seemingly dictatorial position over most of the earth's creatures.

Humans did not choose the path to *Homo dominus*, it chose us. The evolution of the human species has followed basic Darwinian rules. The best-adapted individuals have reproduced their unique characteristics and abilities according to the principle of natural selection. As the generations passed, genetic variants continued to emerge, often simply as a hedge against a changing world, until a recognizably human form appeared. According to current paleontological thinking, human evolution traces some five to seven million years back to an apelike ancestor common to both chimpanzees and humans. Over the ensuing four hundred thousand generations or so, the hominids and chimpanzees diverged to produce two very different species. The chimpanzees endured in their rich forest world, while the hominids asserted control over the larger diversified world.

Can we say that *Homo dominus* has been successful? We are still here, which is about all the success that natural selection can grant. Will we continue to evolve? Probably. Can we determine how? Probably not. If the earth's climate remains stable and the world's pathogens do not mutate into horrific forms, we might predict the course of human evolution. But our planet is not constant, and its pathogens and predators do not recognize human agendas. Other successful species, like the dinosaurs, existed for hundreds of millions of years, and then went extinct. Permanence is not guaranteed to any species.

In the first chapter of this work, I juxtaposed the vision of *Homo sapiens* with the reality of *Homo dominus*. While we cannot predict where *Homo dominus* will eventually go, we can ask how we might become *Homo sapiens*. In part, the answer lies in how we handle this new image of ourselves: that we exist simply to control our world. Some concluding thoughts follow.

The Paleontology of Control

At some point in the distant past, the common ancestor of hominids and chimpanzees lived in small communities in a vast rain forest ecosystem. Over time, according to theory, climate changes caused the ancient habitat to retreat periodically, leaving some ancestral ape species stranded on the margins, while others seized and held the choicest territories in the deep forest. The marginal apes evolved into hominids, then into humans, while the deep forest apes evolved into modern chimpanzees and bonobos, not much different from the common ancestor, according to many paleontologists. Selective pressures for a control strategy were apparently stronger on the margins than in the deep forest. Understanding how this ancient environment generated the critical mix required to build human beings is a key to understanding our origins.

In comparison to other species, the uniqueness of the human control strategy is more quantitative than qualitative. Other species display impressive control behavior in selected aspects of their lives. Arguably, any species with a reasonable amount of cortical mass, plus the requisite subcortical support structures, should exhibit control behavior as part of their repertoire. For example, watching the alpha chimpanzee systematically bullying his subordinates and promoting his allies, we could conclude that he is indeed controlling the group's socio-political structure in order to promote his own personal expectations. Unfortunately, while he might succeed in this particular control task, he seemingly does little else to control the forces affecting his group. He does not manage the long-term food supply, or improve foraging efficiency, or enhance the safety of infants, including his own. He seems utterly incapable of dealing diplomatically with chimpanzees outside his home group, treating them as implacable foes, even though alliances could bring more reproductive opportunities and help safeguard his territory against encroachment by more distant groups. Thus, while the chimpanzee clearly has elements of a control strategy, it seems limited to a few select areas. So, to say that *only* the human species has the capacity to control events overstates the case, but to say that human control exceeds that of other species certainly does not.

To illustrate the differences, try imagining what would happen if a group of humans replaced the chimpanzees in a particular patch of rain forest. What would we see? After a few tense moments of watching and waiting, ever alert to possible threats, the human group would probably start a systematic exploration to inventory resources and assess threats in the area. An encampment would be set up near fresh water. Gatherers would fan out and start transporting food and materials back. Hunting parties would fan out to find game. Craftspeople would gather raw materials and set up their factories. Wisps of smoke from campfires would soon percolate up through the trees. Nursing females, mothers with toddlers, preadolescent children, and the elderly would organize the camps and establish a communal nursery to protect and nurture the young. Throughout it all, we would hear the constant burbling of conversation, rhythmic and musical, in contrast to the hoots and screams of the other apes. In short, with humans substituted for chimpanzees, we would see a very different picture of primate life in the same patch of forest. Instead of random foraging, we would see planning. Instead of mob behavior, we would see organization. Instead of volatile bullying, we would see cooperation.

Humans do not simply accept the environment as it is, they impose their expectations of what the environment needs to be. While they can be as reactive and opportunistic as chimpanzees, they can also plan and manipulate to extract greater value from the resources at hand. Human groups function as intelligent control machines, systematically advancing a shared set of expectations. Chimpanzees, in contrast, function as free agent mobs, sometimes cooperative and always alert to opportunities, but hopelessly vulnerable to selfish bickering. The chimpanzees built their survival strategy around harvesting the seemingly endless bounty of the rain forest, so they could afford to leave much to chance. The hominids built theirs around coping with the rigors of life on the forest margins, so they could not afford to leave anything to chance.

In the addition to asking where the human strategy originated, we need to ask when. Most paleontologists are cautious on the question of humanness in the earliest hominids. The Australopithecines, for example, emerged some three to four million years ago, but in the eyes of most paleontologists, only their enigmatic shift to bipedalism qualifies them for consideration in the human family. Otherwise, they were apelike. Only with the later emergence of genus *Homo* does the human profile clearly emerge: increasing brain size, stone tools, bipedal efficiency, and so on. Some authorities even question the humanity of the earliest forms of genus *Homo*, preferring to bestow the trait of humanness only on those recent ancestors recognizable as *Homo sapiens*.

The question remains, then, is humanness new or old? Obviously, it depends on one's definition. The present work suggests that we define humanness by the ascendancy of control behavior. Our apelike ancestors started on the path to humanness when they developed expectations of how the world should be, and then acted to bring their perceptions into line with those expectations. In this view, humanness is a behavioral distinction, encoded in the neurophysiology of the evolving brain. This behavioral path could have started far back in antiquity, long before the bones and teeth took recognizable hominid shapes. The behavioral origins of control may be more difficult to find than fossil bones and chipped stones, but in the final analysis, this is where the first real stirrings of humanness may lie.

Some paleontologists believe that bipedalism represents the best tangible evidence of the path to humanness. This view pushes our ancestry well back in evolutionary history, probably long before the Australopithecines. Bipedalism was a curious development, however. Paleontologists have struggled with its evolutionary rationale, some playfully calling it "insane." It makes more sense, however, when viewed not as an isolated event, but as one of several concurrent changes in the ancestral hominids, specifically: (1) altered brain function and development; (2) secondary altriciality; and (3) bipedal locomotion. This co-evolutionary model, which I call the Hominid Trinity, suggests that bipedalism was an adaptive response to increasing infant helplessness caused by critical changes in brain development. Chapter 5 makes an argument for this hypothesis. Building the scientific case will require much further investigation.

Once efficient upright locomotion became established, it gave the emerging hominids the ability to transport objects efficiently from place to place, which, in my view, clearly reflects control behavior rather than reactive opportunism. The simple act of moving things from one place to another, repeatedly and deliberately, implies internal expectations of where things *ought to go*. Acting on such expectations brought about a rudimentary transportation economy among the early hominids. This represented a significant shift away from the location-based economy of the ancient apes. With mobility came a system of intrinsic values for objects and commodities. All of these developments accelerated the divergence of the hominids from their apelike ancestors.

In the end, the question of humanness will always be subjective. If we could go back in time and send researchers like Jane Goodall into the ancient habitats, what evidence of humanness would they see in the ancestral apes? They might observe small bands of clever apes, wary and deliberate, perhaps quieter and less volatile than the chimpanzees, moving upright from place to place, carrying food, material, and infants. Dominant males might assert themselves from time to time, but more often, the social order would reflect a greater degree of egalitarian cooperation. Males might stay in closer proximity to particular females, remaining attentive and protective, frequently playing with infants and children, both their own and those of allies and relatives. The surrounding habitat would probably show signs of transformation, like the systematic rearrangement of features or the deliberate placement of objects. In time, observers might be able to tell immediately by sound, sight, and smell when they had moved out of the ape's range and into the hominid's, even though the animals themselves might still bear a physical resemblance.

While we will probably always have scant clues about when humanness first began, we surely will have no shortage of opinions on when it arrived full force. Some will argue that stone tools define the moment. Others will cite dental adaptations or enlarged cranial capacity. Still others might hold out for syntactic language or some indication of consciousness or advanced culture. My preference, simply because it seems so extraordinary, is the migration of *Homo erectus* out of Africa. The fossil evidence suggests they left as early as 1.5 million years ago and traveled all the way to Java, a distance of some ten thousand miles, through widely varying terrain, climates, predators, food supplies, and entirely new populations of natural parasites and pathogens. This remarkable trek reveals a species committed to controlling its world rather than simply reacting to it. These people did not hunker down in the forest and wait for the trees to bear fruit. They packed up their tools and traveling rations and stepped into unknown territory, not knowing what they might find, but apparently confident they could handle anything. We might have recognized them as they passed.

Whatever event you choose as indicative of the human arrival, understand that its precursors probably date back much further, residing in decidedly apelike creatures who probably did not inspire much confidence in the human future. Understanding their contributions to human origins means understanding their subtle variations and adaptations to particular environments. We may find it hard to accept that the first flicker of human awareness occurred inside an apelike cranium, and that it endured not because it made those creatures human, but because it made them better apes. In the beginning, life as a human was less important than survival as an ape. As the generations passed, and the species continued to evolve, the defining traces of that early repertoire crystallized as a recognizable constellation of human traits.

The Future of Control

As an evolutionary driver, control will continue to have implications for human development, both present and future. Humans will always seek control, be it a lone individual living in the wilderness or a population of millions inhabiting a modern city. Achieving control is the principal motivation both for our rich repertoire of cooperative behavior and for the appalling atrocities that we perpetrate on one another. Understanding human behavior simply cannot proceed without identifying the control agendas operating in any situation.

In Chapters 5 and 6, I outlined two powerful manifestations of the human control drive, technology and aggression. By our definition, these are separate traits. Each represents an evolutionary module that extends the rudimentary activities of our ancient ancestors into greatly exaggerated form. Over the generations, however, human technology and aggression have become intertwined. Aggression rewards technological innovation and technology in turn enables more effective aggression. This interplay represents the defining dynamic of modern society. Together, they manifest the push toward unbridled control that has seemingly propelled the human species to the brink of self-annihilation. Be it military weaponry, such as nuclear devices and biological agents, or industrial competition with its attendant pollution and resource depletion, human control has created conditions that could send our species tumbling into oblivion. Like the dinosaurs extinguished by a chance asteroid, humans could succumb to a self-made cataclysm of nuclear, chemical, biological, or thermal poisoning that our fragile bodies cannot withstand and our reproductive rates cannot overcome.

Human society did not reach its current state by collectively deciding to go there. It came about as the aggregate of millions of individual decisions, each made by a person trying to control some aspect of his or her life. We make such decisions based on the information at hand. Sometimes it is clear and unambiguous, other times, fuzzy and uncertain. When the available data do not support rational control, we do not wait for clarity. We simply synthesize imaginary data and construct new hypotheses to fill in the gaps. Myopia regarding facts rarely inhibits the assertion of control. We act to bring our perceptions, however flawed or incomplete, into line with our expectations, however shortsighted or untenable. The outcome, good or bad, creates a new control problem, which demands new actions. The entire process repeats endlessly. The constant in human activity is not intelligence, logic, or truth. It is the attempt to gain and maintain control.

This perpetual striving explains why human societies sometimes lapse into extreme conflict. Unlike most other primates, humans have an enormous capacity for lethal violence within the species. While some chimpanzee groups appear to have the same propensity, they practice it on a considerably reduced scale, without the aid of advanced technology, and then perhaps only as a response to extreme ecological stress. Humans, on the other hand, commit and condone homicide with ease and regularity. Aggression is a natural manifestation of our control strategy. As individuals pursue their personal agendas, they inevitably see others as obstacles to be removed. If the dispute cannot be resolved by nonviolent means, then violence becomes a permissible action. In some cases, it may escalate to acts of extraordinary cruelty, as atrocity and genocide are well within our behavioral repertoire. Indeed, the question posed previously in this book is not why we are prone to lethal violence, but why it ever abates. The drive for control, aggregated over millions of practitioners, creates a momentum not easily turned. Our violent past defines our present; our violent present shows us our future. While we might wish for peace, it will not come if the asking price is the abandonment of control.

Given the increasing technological prowess of human beings, the reduction of aggression and the moderation of conflict will be increasingly critical to our future. Human evolution offers some veiled visions of a peaceful existence. Our predispositions toward altruism (Chapter 7) and cultural exchange (Chapter 8) provide counterbalancing forces against unchecked aggression. While they clearly cannot prevent the rush to war or the indulgence of lethal technologies, they do provide a kind of sea anchor restraining our drift toward self-annihilation. Of course, given the power and reach of modern weapons, even the slightest miscalculation in the coming generations could send us over the brink.

The Case for Altruism

Altruism offers one hope for checking the negative consequences of aggression in the pursuit of control. Altruism, however, is a difficult sociobiological problem. Social scientists suggest that limited forms of altruism, such as kinship and reciprocity, are evolutionarily stable, but the ideal of unconditional self-sacrifice extending throughout society appears biologically untenable. Unless there is some payback to the supposed altruist, the trait invariably drops out of the genetic repertoire under the pressure of selfishness and aggression.

On the other hand, however illogical altruism may seem in sociobiological models, it becomes critically important if the consequence of unchecked aggression is the extinction of the species. Constant warfare depletes reproducing populations; genocide depletes genetic diversity. Both violate basic evolutionary principles. Such egregious behavior may lead to replacement by genetic variants more capable of moderating their aggressive behavior. If the sociobiological models include extinction as the inevitable consequence of selfishness and aggression, then true altruism might emerge as a valid force for moderating human behavior against its own self-destructive instincts.

The fact that most modern hunter-gatherer groups practice altruism suggests that it can endure in human populations (Boehm, 1999). If it restrains selfishness and aggression even slightly, then it may help even the most aggressive human societies avoid self-extinction. While altruism will never completely neutralize aggression, it may help individuals moderate their personal behavior. Propagated across groups, the aggregate of such incremental resistances may stabilize the society while preserving the individual pursuit of control.

If altruism restrains the excesses of *Homo dominus*, does it then define the path to *Homo sapiens*? Will wisdom and justice spring from selfless altruism, or will they come from the heavy hammer of extinction smashing down on the unwise and unjust? The answer is probably both. As *Homo dominus*, we cannot escape our controlling natures, and for that, we will always pay a price in human blood. But if we had not become *Homo dominus*, we might never have survived the harsh realities of the late Miocene. We walked an evolutionary tightrope back then, as indeed we do now. On one side is the intractable violence of control, on the other, extinction as an ape unable to cope with life's challenges. That we still survive means that we have found some source of balance, one manifestation of which may be enduring altruism.

The Case for Culture

While altruism may have helped the human species escape the violence of control, it alone cannot explain the remarkable structure of human society. Indeed, if we relied on altruism as our only protection against ourselves, we probably would still be living in small tribes of hunter-gatherers, perhaps a hundred to two hundred individuals, jealously protective of our territories, and xenophobic in the extreme. Some human groups still live this way, but most reside in enormous societies comprising many thousands of groups intermingled in relative harmony and cooperation. This pattern is unique among primates. Its origin derives in part from the unique invention of human culture.

In chapter 8, I suggested that culture has two distinct components: (1) the transmission of acquired knowledge *within* natural human groups; and (2) the expression of acquired signals, often trivial and biologically meaningless, to moderate aggression *between* groups. The interplay of these two cultural processes defines the unique structure of modern human society.

The first component refers to functional knowledge retained and propagated within the social system I call the sexual/kinship/inlaw/friendship (SKIF) group. Such groups comprise a diverse mixture of pair-bonded couples, their offspring, their extended families and inlaws, plus selected unrelated associates and allies who receive the benefits of kinship. SKIF groups balance the principles of inclusive fitness and genetic diversity in order to achieve evolutionary stability. Within the SKIF circle, humans display extraordinary cooperation, seemingly functioning as a single intelligent organism with a single purpose. This talent reflects the unique cognitive control architectures of our brains, incorporating a powerful predisposition to synchronize with others in a cooperative whole.

Human linguistic communication facilitates this synchronization through its unique syntactic properties. Syntax gives voice to the brain's underlying control structure. It provides a mechanism for synchronizing control systems in separate brains in order to bring greater force and knowledge to shared problems and tasks. When an individual can communicate his or her perceptions, expectations, and intended actions, all other individuals in the group acquire the ability to act on the same basis. In such a system, the knowledge level of the aggregate group has the potential to ratchet up to the level of the smartest, most perceptive, and most experienced individuals in the group. With shared perceptions and expectations, cooperative action is a natural outcome, even among otherwise independent free agents. Human control systems have an inherent ability to acquire and transmit knowledge in this way, giving all members a better chance to control their individual lives.

Equally important, in a stable cultural system, learned knowledge propagates from one generation to the next. Human children have an enormous capacity for observing and imitating the behavior of parents and peers. Arguably, this represents one of the fundamental evolutionary rationales for delaying human brain development into postnatal life, as suggested in Chapter 5. Repeated across the generations, the transfer of learned information raises the group's collective knowledge not only to the highest levels available within the existing group, but to the highest levels attained within the living memory of the group. The steady growth of knowledge makes each generation more capable of controlling its world and in turn passing that capability to subsequent generations. The passage of learned knowledge down through the generations of an enduring SKIF community is a defining feature of human life.

But human cultural development does not stop there. In Chapter 8, I suggested that human societies have developed a second major mode, cultural signaling, that is unique among the ape species. In the chimpanzee cultural model, knowledge accumulated in the local group remains trapped within that group, unable to propagate to other groups due to xenophobic barriers. In such systems, each group must reinvent the wheel, literally. The humans found another way.

Cultural signaling represents a unique and innovative form of between-group communication. Individual SKIF groups may adopt a variety of elaborate external displays that indicate their identities and cultural origins. In early SKIF societies, such displays probably heightened xenophobic tensions. However, over time, certain genetic variations led to the relaxation of strict rules on permissible social displays within a group. Over time, this variant species began to include display elements borrowed from other groups, including their traditional SKIF competitors. Under this cultural signaling hypothesis, these diversified displays, incorporating elements both from the traditional SKIF culture and reflected elements drawn from outside SKIFs, acted as threat-reducing gestures which signaled the sender's intent to behave peacefully and cooperatively with the receiver. When reciprocated between groups, such cultural exchanges masked traditional group identities to create loose, but still relatively stable, social communities. Such extended aggregations gave rise to the ultra-sociality characteristic of human societies. This, in turn, conveyed considerable adaptive benefit to the enculturated groups. Larger group size had the advantages of greater safety against predators and greater concentration of economic, social, and reproductive resources. Moreover, in this shared environment, people learned from one another regardless of their SKIF origins, effectively ratcheting up the general knowledge base to that of the smartest, most innovative contributors across the entire population. Such shared knowledge accelerated the spread of technological innovations and process improvements across the participating groups.

Cultural signaling also explains why so much of what we observe in human culture seems to have little or no biological relevance or perceivable adaptive value. Traditions such as body decoration, clothing styles, architectural design, linguistic dialects, forms of entertainment, and so on, seem to arise almost as random events, perpetuated by particular groups seemingly without adaptive benefit or biological rationale. It is simply "our way," they say. From a social scientific perspective, understanding these seeming trivialities of human culture may be the key to understanding some of human society's deeper profundities.

Cultural signaling may reveal the path that *Homo dominus* must take to become *Homo sapiens*. It says that cultural exchanges offer our best, and perhaps only, hope for escaping the trap of xenophobia. Human history shows that aggression, atrocity, and genocide can be avoided by signaling across the xenophobic boundaries that divide us. If we, as *Homo dominus*, aspire to the wisdom of *Homo sapiens*, we must accept that such cultural exchanges may offer the only proven antidote to lethal conflict. If it can successfully merge thousands of ancient warring bands of *Homo dominus* together into nation-states of great wealth and power, then it certainly offers the promise for preventing future wars and their attendant excesses. A wise species would let this process work, indeed encourage it. Unfortunately, at this writing, the impetus for cultural exchanges seems to have been lost in many places and a resurgence of xenophobic aggression appears to be in full swing.

Even with cultural signaling, violent conflict will always threaten *Homo dominus*. Aggression is a powerful instinct of much longer evolutionary standing than cultural signaling. Like altruism, cultural signaling represents only a brake on aggression, not a true roadblock. Control is the central theme of human nature, and humans will always fight to retain it. Failure to understand this basic instinct guarantees perpetual conflict and the absence of enduring peace.

The Case for Religion

Some see religion as the best path for transforming *Homo dominus* into *Homo sapiens*. Religion is an extraordinary human activity. Some argue that it is simply a cultural phenomenon, like hairstyles or string quartets. However, this seems a gross understatement. Religion is pervasive across human societies, finding expression in almost every human group, from giant nation-states to the most basic hunter-gatherer tribes. While religions may take many diverse forms, which suggests strong cultural influences, the almost universal presence of religion in human life argues for a deep biological foundation in the species. In Chapter 10, I suggested that religion represents one major cultural expression of the biological predisposition toward *spirituality*.

It is important that you understand this concept of spirituality, whether or not you agree with the use of the term. I define it as the human predisposition to deal with the unknown by promulgating hypotheses that somehow convey a sense of control. This is a uniquely human trait because only humans can comprehend that there is an unknown. Other species remain unaware. The theory of *Homo dominus* says that humans are always driven to control events, but not all events are controllable. Death, sickness, predation, climate change, and so on, are events which humans can perceive but which we cannot reliably control. The resulting intractable dissonance drives perpetual, but futile, striving. It is in this context that our spiritual predisposition emerges. This cognitive module instinctively fashions hypotheses that give us the illusion of control. With this illusion established, we proceed with elaborate behavioral programs that reflect our imagined control.

I use the term *superstition* to describe the behavior that emanates from spiritual hypotheses. This term may seem pejorative, but that is not my intent. Virtually everyone engages in superstitious behavior, often without being aware of it. All modern religions build on spiritual hypotheses about the afterlife and the entities that oversee it. In the absence of proof, the religious behavior based on these hypotheses falls within my definition of superstitious behavior.

Modern science also builds on the human spiritual foundation. Indeed the scientific method explicitly sanctions hypothesis-building and its attendant experimental behavior in much the same way that religion does. The history of science also records many instances in which supposedly rational, intelligent scientists behaved irrationally and superstitiously for many years under the power of wrong hypotheses. Viewed in this light, religion and science are the sibling offspring of a deeper spiritual drive. They have both survived over evolutionary time because they both help the controlling human mind find a path through the unknown.

Of course, there are some essential differences between religion and science. Religions deal with really big unknowns, like death and the afterlife. They build elaborate hypotheses about how to control the consequences of death, once we accept that we cannot control death itself. Under these hypotheses, particular behaviors, carried out faithfully, presumably guarantee an individual's passage into the afterlife, according to the particular religious doctrine.

Some scientists consider this anathema, but science cannot compete with religion until it too can deal with death and its consequences. Medicine and biotechnology carry the implied promise of controlling death, but they are rather short on guarantees. While each new procedure or drug that prevents or reverses some life-threatening condition contributes to the aggregate belief that humans will someday achieve immortality, few scientists see this outcome as certain or imminent. That leaves religion as the only source of answers about death and the life after. Religion will therefore exist in human culture as long as human spirituality remains unfulfilled by the answers of science.

While modern science may have difficulty accepting religion as a valid answer to fundamental human needs, the pragmatic *Homo dominus* has no such qualms. Our species has invented and propagated religions in myriad forms. Arguably, their net impact has been positive. They promote stability in human cultures by promulgating convincing hypotheses about the consequences of selfish, immoral, or aggressive behavior. Religions are often the only sustained voices for altruism and peaceful coexistence among cultures otherwise dedicated to selfishness and war. Of course, they are far from perfect. Indeed, some religions have been co-conspirators in aggression, atrocity, and genocide. However, in general, I think we can fairly say that religion has been a stabilizing force in human development and evolution.

What role will religion play in our evolution from *Homo dominus* to *Homo sapiens*? Will *Homo sapiens* believe in God, or gods, as strongly as *Homo dominus*? As a human cultural institution, religion carries all the flaws of other human social systems, but as a way of dealing with the unknown, it comforts us with its confident hypotheses. Perhaps the path to *Homo sapiens* lies in the inclusive view that we take wisdom, sensibility, and justice from whence it comes. If religion makes us wise, we should welcome it, regardless of whether it has scientific credibility. If aspiring to become *Homo sapiens* means moderating our controlling natures, perhaps the faithful practice of goodwill religions offers the only viable path.

The Case for Science

Science, like religion, also offers us the promise of conquering the unknown. Unlike religion, however, science works best by seeking to destroy the very hypotheses it builds. It does this by systematically gathering and arranging perceptions (facts) to match or mismatch the expectation (hypothesis) at hand. In applying such discipline, science differs from religion. The scientific mind accepts hypotheses as sustainable truths only insofar as the facts justify, while the religious mind often lapses into preserving cherished hypotheses against the emerging facts. Science grew out of the need to improve control in certain areas of human life by substituting discipline and rationality for religion's faith and superstition. In these areas, science offers the most pragmatic approach invented by the human mind for dealing with the unknown.

Unfortunately, scientific institutions sometimes display the same flaws as religious institutions. While the ideal of science is truth, the reality is control. Modern science, with its focus on technological manipulation, often promises control that it cannot reliably deliver. We even see pronouncements that science has carried the human species beyond normal evolution, that natural selection has given way to cultural selection. According to this view, science can now control the very process that gave us our ability to control. With further research and technology, the speculation goes, we will have the power to redesign ourselves according to our own vision of ourselves: healthy, immortal, and always in control.

Nature may surprise us on this one. Despite our technical advances, our knowledge remains well short of what we would need to impact human evolution. While we can achieve isolated results, which the beneficiaries might consider miraculous, the general state of human evolutionary fitness probably has not changed much over the past few millennia. Science has accomplished some noteworthy improvements in health, but few of these represent sustainable improvements in genetic fitness. Indeed, some would argue, perhaps harshly, that modern medicine has actually reduced the genetic fitness of the human species against genuine evolutionary challenges. Individuals who would surely have perished in the past now reproduce successfully under the modern medical shield. For example, I wear glasses. If I were transported back in time to the ancient savanna, *without* my corrective lenses, I would have almost no chance of survival even if I knew what to do. I would be effectively blind. Only the altruism of the indigenous *Homo erectus* could save me from the fatal affliction of myopia.

Recent advances in genetic cloning may represent another example of technology substituting for wisdom. Cloning technology has given us better control over certain aspects of biological reproduction by eliminating the annoying problem of genetic diversity. But is that really wise? Experience from the so-called green revolution illustrates the problem (Lappé, Collins, and Rosset 1998). Introduced in the 20th century, food crops based on strains with much less genetic diversity than natural varieties took over commercial agriculture with the promise that they would produce higher yields at less cost. The experiment stumbled, however, when scientists realized that reducing genetic diversity made these crops more susceptible to parasites and disease, thus compromising the expected benefits. To combat this vulnerability, farmers had to increase their use of pesticides, divert greater quantities of water to the fields, and sacrifice natural vegetation. As a result, total food production did actually increase but at the cost of a greater load of biological toxins in the environment and the loss of habitat for other species, including many beneficial plants, animals, and insects. Thus, cloning technology gave us better control of one element of the biological equation, but sacrificed control in others. If future human health deteriorates because of the accumulation of toxic waste and the destruction of beneficial ecosystems, what have we really gained?

Even more problematic, the promise of technological control often obscures cultural deficiencies that exacerbate biological problems. In the green revolution, total food production increased, yet in many places, hunger remained chronic. Some estimates suggest that malnutrition threatens nearly 800 million people worldwide (Rosset, Collins, and Lappé 2000; Pretty 1998). This happens because transportation and distribution systems remain inadequate in many areas. Grain rots in warehouses while people go hungry, simply because local governments cannot or will not transport and distribute the resources properly. Technological advances in food production have no benefit if the products cannot reach the people in need. Spending less on technical wizardry and more on basic transportation, distribution, and economic infrastructure might have a greater impact.

Misguided cultural agendas can also escalate natural problems into potential catastrophes. The eruption of the HIV epidemic in the 1980s became a biological catastrophe. Yet, to witness the initial response of our cultural institutions, one might have imagined that HIV was simply a minor nuisance over which we would quickly gain control. Perhaps the optimistic promises of medical technology had lulled us into complacency about disease. As of 2007, according to estimates from the United Nations (UNAIDS) and the World Health Organization (WHO), nearly thirty-one million adults and three million children live with HIV infection. Over twenty-five million have died since 1981. This is considerably more than a nuisance. Moreover, the history of the epidemic suggests that better judgment at the beginning might have contained and perhaps even prevented it (Shilts 1987). The hubris of technological control and misguided cultural agendas apparently overwhelmed biological truth and needlessly unleashed a plague on the world.

Would *Homo sapiens* do a better job than *Homo dominus* in dealing with such problems? Wisdom and justice say we should address human problems from a starting point of truth, without regard to cultural expediency or technological hubris. The latter clearly reflect the legacy of *Homo dominus*. The hallmark of that species has been successful control through the intelligent application of resources. However, we cannot sustain intelligence if we cannot distinguish truth from noise. In the end, if we continue to value control above truth, we will continue to fall short of the promise of *Homo sapiens*.

Accepting Ourselves

Most humans do not really understand or accept themselves as a biological species. We gather facts about ourselves; indeed no species studies itself as we do. But without a unifying theme that tells us what the facts mean, we keep sorting and re-sorting them into various unconnected piles. The questions keep recurring: What does humanness really mean? Who are we? What are we? We want to believe that the name *Homo sapiens* truly applies, that we behave with wisdom, sensibility, and justice, but the facts suggest otherwise. We have actually become *Homo dominus*, not a malevolent end by any means, simply a creature self-absorbed with controlling its world. If this characterization fits the facts, if it captures the human definition, then we must accept it. We may not like this portrait of ourselves. If not, our only option is to moderate our controlling natures, probably a futile exercise and unlikely to bring swift results.

Accepting ourselves as a controlling species may give us a better understanding of how human institutions work, and perhaps why they fail. We cannot escape what we are, but we can factor it into our assessment of problems and solutions. When facing difficult problems, some Christians ask, "What would Jesus do?" In a secular context, perhaps we could amend that to "What would *Homo sapiens* do?" Whether *Homo dominus* can accept the answer is another question.

Control is not bad or good. It is simply what we do. Either it works or it does not. The ultimate arbiter of biological strategy is natural selection, and its decisions are not always immediately discernable. All we know is that several million years ago, at the edge of a vast rain forest, a few scattered bands of frightened, hungry apes were compelled to take control of their future rather than waiting for things to happen. That spark set off a chain reaction that propelled our species out of apedom and into a position of planetary power. It happened through the simple mechanism of control and, for the moment at least, we seem to have achieved a state of invulnerability unimagined by those ancient ancestors. Whether we have the wisdom to maintain it against the nature's future challenges remains an open question. Nature always makes things uncertain for life on earth.

* * *

REFERENCES

Aiello, L. C., and P. Wheeler. 1995. The expensive tissue hypothesis: The brain and the digestive system in human and primate evolution. *Current Anthropology* 36:199–221.

Aiken, N. E. 1998. *The Biological Origins of Art.* Westport, CT: Praeger.

Alexander, R. D. 1974. The evolution of social behavior. *Ann. Rev. Ecol. Syst.* 5:325–383.

Alexander, R. D., and D. W. Tinkle. 1968. Review of On Aggression, by Konrad Lorenz, and The Territorial Imperative, by Robert Ardrey. *Bioscience* 18:245–248.

Alexander, R. D., and K. M. Noonan. 1979. Concealment of ovulation, parental care, and human social evolution. In *Evolutionary Biology and Human Social Behavior: An Anthropological Perspective*, ed. N. A. Chagnon and W. G. Irons, 436–453. North Scituate, MA: Duxbury Press.

Altmann, J., S. C. Alberts, S. A. Haines, J. Dubach, P. Muruthi, T. Coote, E. Geffen, D. J. Cheesman, R. S. Mututua, S. N. Saiyalel, R. K. Wayne, R. C. Lacy, and M. W. Bruford. 1996. Behavior predicts genetic structure in a wild primate group. *Proc. Natl. Acad. Sci. USA* 93:5797–5801.

Amaral, L. Q. 2007. Mechanical analysis of infant carrying in hominoids. *Naturwissenschaften* 95:129–142.

Aunger, R. 2002. *The Electric Meme: A New Theory of How We Think.* New York: Free Press.

Aunger, R., ed. 2000. *Darwinizing Culture: The Status of Memetics as a Science.* Oxford: Oxford University Press.

Axelrod, R. 1984. *The Evolution of Cooperation.* New York: Basic Books.

Axelrod, R., and W. D. Hamilton. 1981. The evolution of cooperation. *Science* 211:1390–1396.

Bainbridge, D. 2008. *Beyond the Zonules of Zinn: A Fantastic Journey Through Your Brain*. Cambridge, MA: Harvard University Press.

Bates, D. G., and F. Plog. 1990. *Cultural Anthropology*. New York: McGraw-Hill.

Beauregard, M., and D. O'Leary. 2007. *The Spiritual Brain: A Neuroscientist's Case for the Existence of the Soul*. New York: HarperCollins.

Begun, D. R. 2003. Planet of the apes. *Scientific American* 289: 74–83.

Bell, A. H., M. A. Meredith, A. J. Van Opstal, and D. P. Munoz. 2005. Crossmodal integration in the primate superior colliculus underlying the preparation and initiation of saccadic eye movements. *J. Neurophysiol.* 93:3659–3673.

Benshoof, L., and R. Thornhill. 1979. The evolution of monogamy and concealed ovulation in humans. *J. Soc. Biol. Structures* 2:95–106.

Berard, J. D., P. Nurnberg, J. T. Epplen, and J. Schmidtke. 1993. Male rank, reproductive behavior and reproductive success in free-ranging Rhesus Macaques. *Primates* 34:481–489.

Berard, J. D., P. Nurnberg, J. T. Epplen, and J. Schmidtke. 1994. Alternative reproductive tactics and reproductive success in male Rhesus Macaques. *Behaviour* 129:177–201.

Berggren, W. A. and J. A. Van Couvering, eds. 1984. *Catastrophes and Earth History: The New Uniformitarianism*. Princeton: Princeton University Press.

Bernstein, I. S. 1981. Dominance: The baby and the bathwater. *Behav. Brain Sci.* 4:419–429.

Berra, Y. 1998. *The Yogi Book: I Really Didn't Say Everything I Said!* New York: Workman Publishing.

Bickerton, D. 1990. *Language and Species*. Chicago: University of Chicago Press.

Bickerton, D. 1995. *Language and Human Behavior*. Seattle: University of Washington Press.

Bickerton, D. 2008. *Bastard Tongues: A Trailblazing Linguist Finds Clues to Our Common Humanity in the World's Lowliest Languages*. New York: Farrar, Straus and Giroux.

Bickhard, M. H. 1993. Representational content in humans and machines. *J. Exp. Theoret. Artif. Intell.* 5:285–333.

Bickhard, M. H. 1999. Interaction and representation. *Theory Psychol.* 9:435–458.

Bischoff, J. L., D. D. Shamp, J. L. Arsuaga, E. Carbonell, and J. M. Bermudez de Castro. 2003. The Sima de los Huesos hominids date to beyond U/Th equilibrium (>350 kyr) and perhaps to 400–500 kyr: New radiometric dates. *Journal of Archaeological Science* 30:275–280.

Blackmore, S. 1999. *The Meme Machine.* Oxford: Oxford University Press.

Blumenschine, R. J., and J. A. Cavallo. 1992. Scavenging and human evolution. *Scientific American* 267:90–96.

Boehm, C. 1999. *Hierarchy in the Forest: The Evolution of Egalitarian Behavior.* Cambridge, MA: Harvard University Press.

Boesch, C., and H. Boesch-Aschermann. 2000. *Chimpanzees of the Tai forest: Behavioral Ecology and Evolution.* Oxford: Oxford University Press.

Bowen, F. ed. 1983. *The Metaphysics of Sir William Hamilton.* Cambridge: Sever & Francis.

Boyer, P. 2001. *Religion Explained: The Evolutionary Origins of Religious Thought.* New York: Basic Books.

Braitenberg, V. 1984. *Vehicles: Experiments in Synthetic Psychology.* Cambridge, MA: MIT Press.

Brodie, R. 1995. *Virus of the Mind: The New Science of the Meme.* Seattle: Integral Press.

Brunet, M., F. Guy, D. Pilbeam, H. T. Mackaye, A. Likius, D. Ahounta, A. Beauvilain, C. Blondel, H. Bocherens, J.-R. Boisserie, L. De Bonis, Y. Coppens, J. Dejax, C. Denys, P. Duringer, V. Eisenmann, G. Fanone, P. Fronty, D. Geraads, T. Lehmann, F. Lihoreau, A. Louchart, A. Mahamat, G. Merceron, G. Mouchelin, O. Otero, P. P. Campomanes, M. Ponce De Leon, J.-C. Rage, M. Sapanet, M. Schuster, J. Sudre, P. Tassy, X. Valentin, P. Vignaud, L. Viriot, A. Zazzo, and C. Zollikofer. 2002. A new hominid from the upper Miocene of Chad, Central Africa. *Nature*, 418:145–151.

Buller, D. J. 2005. *Adapting Minds: Evolutionary Psychology and the Persistent Quest for Human Nature.* Cambridge, MA: MIT Press.

Butler, A. B., and W. Hodos. 1996. *Comparative Vertebrate Neuroanatomy.* New York: Wiley-Liss.

Byrne, R. W., and N. Corp. 2004. Neocortex size predicts deception rate in primates. *Proceedings of the Royal Society B* 271:1693–1699.

Calvert, G. A., C. Spence, and B. E. Stein. 2004. *The Handbook of Multisensory Processes.* Cambridge, MA: MIT Press.

Calvin, W. H. 1996. *The Cerebral Code: Thinking a Thought in the Mosaics of the Mind.* Cambridge: MIT Press.

Calvin, W. H. 2002. *A Brain for All Seasons: Human Evolution and Abrupt Climate Change.* Chicago: University of Chicago Press.

Calvin, W. H., and D. Bickerton. 2000. *Lingua ex Machina: Reconciling Darwin and Chomsky with the Human Brain.* Cambridge, MA: MIT Press.

Cameron, D., and C. Groves. 2004. *Bones, Stones and Molecules: "Out of Africa" and Human Origins.* New York: Elsevier.

Carter, C. S., and E. B. Keverne. 2002. The neurobiology of social affiliation and pair bonding. In *Hormones, Brain, and Behavior, Vol. 1,* ed. D. W. Pfaff, A. P. Arnold, A. M. Etgen, S. E. Fahrbach, and R. T. Rubin, 299–337. New York: Academic Press.

Chalmers, D. J. 1998. *The Conscious Mind: In Search of a Fundamental Theory.* New York: Oxford University Press.

Chapple, E. D. 1970. *Culture and Biological Man.* New York: Holt, Rinehart and Winston.

Cheney, D. L. 1987. Interactions and relationships between groups. In *Primate Societies,* ed. B. B. Smuts, D. L. Cheney, R. M. Seyfarth, R. W. Wrangham, and T. Struhsaker, 267–281. Chicago: Chicago University Press.

Chomsky, N. 1957. *Syntactic Structures.* The Hague: Mouton.

Churchland, P. S. 1986. *Neurophilosophy.* Cambridge: MIT Press.

Cloak, F. T. 1975. Is a cultural ethology possible? *Human Ecology* 3:161–182.

Corballis, M. C. 2003. From hand to mouth: The gestural origins of language. In *Language Evolution: The States of the Art,* ed. M. H. Christiansen and S. Kirby, 201–218. New York: Oxford University Press.

Coss, R. G. 1968. The ethological command in art. *Leonardo* 1:273–287.

Crick, F. H. C. 1994. *The Astonishing Hypothesis: The Scientific Search for the Soul*. New York: Simon & Schuster.

Crystal, D., ed. 1995. *The Cambridge Encyclopedia of the English Language*. Cambridge: Cambridge University Press.

Cziko, G. A. 1995. *Without Miracles: Universal Selection Theory and the Second Darwinian Revolution*. Cambridge: MIT Press.

Cziko, G. A. 2000. *The Things We Do: Using the Lessons of Bernard and Darwin to Understand the What, How, and Why of Our Behavior*. Cambridge: MIT Press.

Daly, M., and M. Wilson. 1978. *Sex, Evolution and Behavior*. North Scituate, MA: Duxbury Press.

Damasio, A. R. 1999. *The Feeling of What Happens: Body and Emotion in the Making of Consciousness*. London: Heineman.

Damasio, A. R., and D. Tranel. 1993. Nouns and verbs are retrieved with differently distributed neural systems. *Proc. Natl. Acad. Sci. USA* 90:4957–4760.

Damasio, H., T. J. Grabowski, D. Tranel, R. D. Hichwa, and A. R. Damasio. 1996. A neural basis for lexical retrieval. *Nature* 380:499–505.

Darwin, C. 1859. *On the Origin of Species by Means of Natural Selection: Or the Preservation of Favoured Races in the Struggle for Life*. London: John Murray.

Darwin, C., and A. R. Wallace. 1858. On the tendency of species to form varieties; and on the perpetuation of varieties and species by natural means of selection. *J. Proc. Linnean Soc., Zool.* 3:45–62.

Dawkins, R. 1976. *The Selfish Gene*. New York: Oxford University Press.

Dawkins, R. 1993. Viruses of the mind. In *Dennett and His Critics: Demystifying Mind*, ed. B. Dalhbom, 13–27. Cambridge: Blackwell.

de Saussure, F. 1966. *Course in General Linguistics*. 1916. Reprint, New York: McGraw-Hill.

de Waal, F. B. M. 2001. Apes from Venus: Bonobos and human social evolution. In *Tree of Origin: What Primate Behavior Can Tell Us About Human Social Evolution*, ed. F. B. M. de Waal, 39–68. Cambridge, MA: Harvard University Press.

Deacon, T. 1997. *The Symbolic Species: The Co-Evolution of Language and the Brain*. New York: Norton.

Delgado-Escueta, A. V., W. A. Wilson, R. W. Olsen, and R. J. Porter, eds. 1999. *Jasper's Basic Mechanisms of the Epilepsies, 3rd Edition*. Philadelphia: Lippincott Williams & Wilkins.

Dennett, D. 1995. *Darwin's Dangerous Idea*. New York: Simon & Schuster.

Dennett, D. C. 1991. *Consciousness Explained*. Boston: Little, Brown.

Dewsbury, D. A. 1982. Dominance rank, copulatory behavior, and differential reproduction. *Q. Rev. Biol.* 57:135–159.

Diamond, J. M. 1992. *The Third Chimpanzee: The Evolution and Future of the Human Animal*. New York: HarperCollins.

Diamond, M. C. 1988. *Enriching Heredity*. New York: The Free Press.

Dissanayake, E. 1992. *Homo Aestheticus: Where Art Comes From and Why*. New York: Free Press.

du Preez, P. 1994. *Genocide: The Psychology of Mass Murder*. London: Marion Boyars.

Dunbar, R. 1988. *Primate Social Systems*. London: Chapman and Hall.

Dunbar, R. 1996. *Grooming, Gossip, and the Evolution of Language*. Cambridge, MA: Harvard University Press.

Dutton, D. 2009. *The Art Instinct: Beauty, Pleasure, and Human Evolution*. New York: Bloomsbury Press.

Eades, M. R., and M. D. Eades. 2001. *Protein Power Lifeplan*. New York: Warner Books.

Edelman, G. M. 1992. *Bright Air, Brilliant Fire: On the Matter of the Mind*. New York: Basic Books.

Edelman, G. M. 2006. *Second Nature: Brain Science and Human Knowledge*. New Haven, CT: Yale University Press.

Eldredge, N. and S. J. Gould. 1972. Punctuated equilibria: An alternative to phyletic gradualism. In *Models in Paleobiology,* ed. T. J. F. Schopf, 82–115. San Francisco: Freeman, Cooper.

Ellis, L. 1995. Dominance and reproductive success among nonhuman animals: A cross-species comparison. *Ethol. Sociobiol.* 16:257–333.

Enard, W., P. Khaitovich, J. Klose, S. Zöllner, F. Heissig, P. Giavalisco, K. Nieselt-Struwe, E. Muchmore, A. Varki, R. Ravid, G. M. Doxiadis, R. E. Bontrop, and S. Pääbo. 2002. Intra- and interspecific variation in primate gene expression patterns. *Science* 296:340–343.

Erikson, E. H. 1966. *Childhood and Society.* Rev. ed. New York: Norton.

Etkin, W. 1954. Social behavior and the evolution of man's mental faculties. *Amer. Nat.* 88:129–142.

Falk, D. 1983. Cerebral cortices of east African early hominids. *Science* 221:1072–1074.

Falk, D. 1992. *Braindance.* New York: Henry Holt.

Felleman, D. J., and D. C. Van Essen. 1991. Distributed hierarchical processing in the primate cerebral cortex. *Cereb. Cortex* 1:1–47.

Ferrill, A. 1985. *Origins of War: From the Stone Age to Alexander the Great.* New York: Thames & Hudson.

Filler, A. G. 2007. *The Upright Ape: A New Origin of the Species.* Franklin Lakes, NJ: New Page Books.

Fisher, H. 1992. *Anatomy of Love: The Mysteries of Mating, Marriage, and Why We Stray.* New York: Simon & Schuster.

Fleagle, J. G. 1999. *Primate Adaptation and Evolution.* San Diego: Academic Press.

Fodor, J. 1983. *The Modularity of the Mind.* Cambridge: MIT Press.

Foley, R. A. 1987. *Another Unique Species: Patterns in Human Evolutionary Ecology.* New York: Wiley.

Foley, R. A., and P. C. Lee. 1991. Ecology and energetics of encephalization in hominid evolution. *Phil. Trans. Royal Soc. Lond. B.* 334:223–232.

Fox, R. 1980. *The Red Lamp of Incest.* New York: Dutton.

Frazer, J. 1922. *The Golden Bough: A Study in Magic and Religion*. New York: Macmillan.

Frith, C., and R. Dolan. 1996. The role of the prefrontal cortex in higher cognitive functions. *Cogn. Brain Res.* 5:175–181.

Gabrieli, J., R. Poldrack, and J. Desmond. 1998. The role of prefrontal cortex in language and memory. *Proc. Natl. Acad. Sci. USA* 95:906–913.

Gallup, G. G. 1970. Chimpanzees: Self-recognition. *Science* 167:86–87.

Gardner, H. 1983. *Frames of Mind: The Theory of Multiple Intelligences*. New York: Basic Books.

Gardner, H. 1985. *The Mind's New Science: A History of the Cognitive Revolution*. New York: Basic Books.

Gardner, R. A., and B. T. Gardner. 1977. Teaching sign language to a chimpanzee. *Science* 197:664–672.

Gazzaniga, M. S. 2008. *Human: The Science Behind What Makes Us Unique*. New York: HarperCollins.

Gazzaniga, M. S., R. B. Ivry, and G. R. Mangun. 1998. *Cognitive Neuroscience: The Biology of the Mind*. New York: Norton.

Geary, D. C. 1998. *Male, Female: The Evolution of Human Sex Differences*. Washington: American Psychological Association.

Ghiglieri, M. P. 1987. War among the chimps. *Discover* 8:67–76.

Ghiglieri, M. P. 1988. *East of the Mountains of the Moon: The Chimpanzees of Kibale Forest*. New York: Free Press.

Gintis, H., S. Bowles, R. Boyd, and E. Fehr. 2003. Explaining altruistic behavior in humans. *Evolution and Human Behavior* 24:153–172.

Gmelch, G. 1978. Baseball magic. *Human Nature* 1:32–50.

Gmelch, G. 2001. *Inside Pitch: Life in Professional Baseball*. Washington: Smithsonian Institution Press.

Goldberg, E. 2001. *The Executive Brain: Frontal Lobes and the Civilized Mind*. New York: Oxford University Press.

Goldschmidt, W. 1960. *Understanding Human Society.* London: Routledge & K. Paul.

Goodall, J. 1986. *The Chimpanzees of Gombe: Patterns of Behavior.* Cambridge: Harvard University Press.

Gould, S. J. 1976. Human babies as embryos. *Natural History* 85:22–26.

Gould, S. J. 1977. *Ever Since Darwin: Reflections in Natural History.* New York: Norton.

Gowlett, J. A. J. 1992. Tools—the paleolithic record. In *The Cambridge Encyclopedia of Human Evolution*, ed. S. Jones, R. Martin, and D. Pilbeam, 350–360. Cambridge: Cambridge University Press.

Grabowski, T. J., H. Damasio, and A. R. Damasio. 1998. Premotor and prefrontal correlates of category-related lexical retrieval. *Neuroimage* 7:232–243.

Greenspan, S. I., and S. G. Shanker. 2004. *The First Idea: How Symbols, Language, and Intelligence Evolved From Our Primate Ancestors to Modern Humans.* Cambridge, MA: Perseus.

Grimes, B. F., ed. 2000. *Ethnologue.* 14th ed. Dallas: SIL International.

Grush, R. 2003. In defense of some "Cartesian" assumptions concerning the brain and its operation. *Biol. Phil.* 18:53–93.

Haldane, J. B. S. 1955. Population genetics. *New Biol.* 18:34–51.

Hamilton, W. D. 1964. The genetical evolution of social behaviour. *J. Theoret. Biol.* 7:1–52.

Harris, M. 1956. *Bang the Drum Slowly.* New York: Alfred A. Knopf.

Harvey, P. H., and T. H. Clutton-Brock. 1985. Life history variation in primates. *Evolution* 39:557–581.

Hauser, M. D. 1996. *The Evolution of Communication.* Cambridge, MA: MIT Press.

Hausfater, G., J. Altmann, and S. Altmann. 1982. Long-term consistency of dominance relationships among female baboons (Papio cyncocephalus). *Science* 217:752–755.

Hawkins, J., and S. Blakeslee. 2004. *On Intelligence*. New York: Henry Holt.

Hebb, D. O. 1949. *The Organization of Behavior: A Neuropsychological Theory*. New York: Wiley.

Hedges, C. 2002. *War Is a Force That Gives Us Meaning*. New York: Random House.

Heinz, S., G. Baron, and H. Frahm. 1988. Comparative size of brains and brain components. In *Comparative Primate Biology, Volume IV: Neurosciences*, ed. H. S. Steklis and J. Erwin, 1–38. New York: Liss.

Hickok, G., U. Bellugi, and E. S. Klima. 1998. What's right about the neural organization of sign language? A perspective on recent neuroimaging results. *Trends in Cognitive Sciences* 2:465–468.

Hockett, C. F., and R. Ascher. 1964. The human revolution. *Cultural Anthropology* 5:135–168.

Hofstadter, Douglas. 2007. *I Am a Strange Loop*. New York: Basic Books.

Holliday, M. A. 1978. Body composition and energy needs during growth. In *Human Growth. Volume 2: Postnatal Growth*, ed. F. Falkner and J. M. Tanner, 117–139. New York: Plenum Press.

Holloway, R. L. 1983. Cerebral brain endocast pattern of Australopithecus afarensis hominid. *Nature* 303:420–422.

Holloway, R. L. 1996. Evolution of the human brain. In *Handbook of Human Symbolic Evolution, Chapter 4*, ed. A. Lock and C. Peters, 74–116. New York: Oxford University Press.

Holloway, R. L., D. C. Broadfield, and M. S. Yuan. 2001. Revisiting australopithecine visual striate cortex: Newer data from chimpanzee and human brains suggest it could have been reduced during australopithecine times. In *Evolutionary Anatomy of the Primate Cerebral Cortex*, ed. D. Falk and K. Gibbon, 177–186. Cambridge: Cambridge University Press.

Hrdy, S. B. 1981. *The Woman That Never Evolved*. Cambridge, MA: Harvard University Press.

Hubel, D. H., and T. N. Wiesel. 1959. Receptive fields of single neurones in the cat's striate cortex. *J. Physiol.* 148:574–591.

Hubel, D. H., and T. N. Wiesel. 1962. Receptive fields, binocular interaction and functional architecture in the cat's visual cortex. *J. Physiol.* 166:106–154.

Hubel, D. H., and T. N. Wiesel. 1968. Receptive fields and functional architecture of monkey striate cortex. *J. Physiol.* 195:215–243.

Hurley, S. 1998. *Consciousness in Action.* Cambridge, MA: Harvard University Press.

Hurley, S. 2001. Perception and action: Alternative views. *Synthese* 129:3–40.

Hurley, S. 2006. Active perception and perceiving action: The shared circuits hypothesis. In *Perceptual Experience,* ed. T. Gendler and J. Hawthorne, 205–259. New York: Oxford University Press.

Hutchins, E. 1995. *Cognition in the Wild.* Cambridge: MIT Press.

Imanishi, K. 1957. Identification: A process of enculturation in the subhuman society of Macaca fuscata. *Primates* 1:1–29.

Insel, T. R., and L. J. Young. 2001. The neurobiology of attachment. *Nature Reviews in Neuroscience* 2:129–136.

Itzkoff, S. W. 1985. *Triumph of the Intelligent: The Creation of Homo sapiens sapiens.* Ashfield, MA: Paideia.

Iwamoto, M. 1985. Bipedalism of Japanese monkeys and carrying models of hominization. In *Primate Morphophysiology, Locomotor Analyses and Human Bipedalism,* ed. S. Kondo, 251–260, Tokyo: Tokyo University Press.

Jackendoff, R. 1987. *Consciousness and the Computational Mind.* Cambridge: MIT Press.

Jackendoff, R. 2002. *Foundations of Language.* Oxford: Oxford University Press.

Jerison, H. J. 1973. *Evolution of the Brain and Intelligence.* New York: Academic Press.

Johanson, D. C., and M. A. Edey. 1981. *Lucy: The Beginnings of Mankind.* New York: Simon and Schuster.

Johanson, D. C., T. D. White, and Y. Coppens. 1978. A new species of the genus Australopithecus (Primates: Hominidae) from the Pliocene of eastern Africa. *Kirtlandia* 28:1–14.

Johnson, S. H. 2000. Thinking ahead: The case for motor imagery in prospective judgements of prehension. *Cognition* 74:33–70.

Kaas, J. H., and C. E. Collins. 2001. Evolving ideas of brain evolution. Nature 411:141–142.

Keenan, J. P. 2003. *The Face in the Mirror: The Search for the Origins of Consciousness*. New York: HarperCollins.

Kelly, J. 1992. Evolution of apes. In *The Cambridge Encyclopedia of Human Evolution*, eds. S. Jones, R. Martin, and D. Pilbeam, 223–230. Cambridge: Cambridge University Press.

Kenneally, C. 2007. *The First Word: The Search for the Origins of Language*. New York: Penguin Books.

Kingdon, J. 2003. *Lowly Origin: Where, When, and Why or Ancestors First Stood Up*. Princeton: Princeton University Press.

Knauft, B. M. 1989. Sociality versus self-interest in human evolution. *Behav. Brain Sci.*12:712–713.

Koch, C. 1998. *Biophysics of Computation: Information Processing in Single Neurons*. New York: Oxford University Press.

Kosslyn, S. M., and A. L. Sussman. 1995. Roles of imagery in perception: Or, there is no such thing as immaculate perception. In *The Cognitive Neurosciences*, ed. M. S. Gazzaniga, 1035–1042. Cambridge: MIT Press.

Kroeber, A. L. 1928. Sub-human culture beginnings. *Quarterly Review of Biology* 3:328–330.

Lampert, A. 1997. *The Evolution of Love*. Westport, CT: Praeger.

Lappé, F. L., J. Collins, and P. Rosset. 1998. *World Hunger: Twelve Myths*. New York: Grove Press.

Larson, E. J. 1997. *Summer for the Gods: The Scopes Trial and America's Continuing Debate over Science and Religion*. New York: Basic Books.

Leakey, L. S. B., P. V. Tobias, and J. R. Napier. 1964. A new species of genus Homo from Olduvai Gorge. *Nature* 202:7–9.

Leakey, M. D. 1971. *Olduvai Gorge, Volume 3: Excavations in Beds I and II, 1960–1963.* Cambridge: Cambridge University Press.

Leakey, M. D., and R. L. Hay. 1979. Pliocene footprints in the Laetoli beds, northern Tanzania. *Nature* 278:317–328.

Leakey, M. G., C. S. Feibel, I. McDougall, and A. C. Walker. 1995. New Four-million-year-old hominid species from Kanapoi and Allia Bay, Kenya. *Nature* 376:565–571.

Leakey, R., and R. Lewin. 1992. *Origins Reconsidered: In Search of What Makes Us Human.* New York: Doubleday.

LeDoux, J. E. 1996. *The Emotional Brain: The Mysterious Underpinnings of Emotional Life.* New York: Simon & Schuster.

Leonard, W. R., and M. L. Robertson. 1994. Evolutionary perspectives on human nutrition: The influence of brain and body size on diet and metabolism. *Amer. J. Human Biol.* 6:77–88.

Lewin, R. 1993. *The Origin of Modern Humans.* New York: Scientific American Library.

Lieberman, P. 1984. *The Biology and Evolution of Language.* Cambridge, MA: Harvard University Press.

Lieberman, P. 1991. *Uniquely Human: The Evolution of Speech, Thought and Selfless Behavior.* Cambridge, MA: Harvard University Press.

Lieberman, P., E. S. Crelin, and D. H. Klatt. 1972. Phonetic ability and related anatomy of the newborn and adult human, Neanderthal man, and the chimpanzee. *American Anthropologist* 74:287–307.

Linnaeus, C. 1758. *Systema naturae per regna tria naturae, secundum classes, ordines, genera, species cum characteribus, differentiis, synonymis, locis.* 10th ed. Stockholm: L. Salvius.

Llinas, R. 2001. *I of the Vortex.* Cambridge, MA: MIT Press.

Lorenz, K. 1963. *On Aggression.* London: Methuen.

Lovejoy, C. O. 1981. The origin of man. *Science* 211:341–350.

Lumsden, C. J., and E. O. Wilson. 1983. *Promethean Fire: Reflections on the Origin of Mind.* Cambridge, MA: Harvard University Press.

Lynch, A. 1996. *Thought Contagion: How Belief Spreads Through Society.* New York: Basic Books.

Macaluso, E. 2006. Multisensory processing in sensory-specific cortical areas. *The Neuroscientist* 12: 327–338.

Malinowski, B. 1954. *Magic, Science and Religion.* 1948. Reprint, Garden City, NY: Doubleday.

Manson, J. H., and R. W. Wrangham. 1991. Intergroup aggression in chimpanzees and humans. *Current Anthropol.* 32:369–77.

Marken, R. S. 1986. Perceptual organization of behavior: A hierarchical control model of coordinated action. *J. Exp. Psychol. Hum. Percept. Perform.* 12:267–276.

Marken, R. S. 1992. *Mind Readings: Experimental Studies of Purpose.* Gravel Switch, KY: CSG Press.

Martin, R. D. 1984. Body size, brain size and feeding strategies. In *Food Acquisition and Processing in Primates,* ed. D. Chivers, B. Wood, and A. Bilsborough, 73–104. New York: Plenum Press.

Matsumura, M., ed. 1995. *Voices for Evolution.* Berkeley, CA: The National Center for Science Education.

McGrew, W. C. 2001. The nature of culture: Prospects and pitfalls of cultural primatology. In *Tree of Origin: What Primate Behavior Can Tell Us About Human Social Evolution,* ed. F. B. M. de Waal, 229–254. Cambridge, MA: Harvard University Press.

McHenry, H. M. 1994. Tempo and mode in human evolution. *Proc. Natl. Acad. Sci. USA* 91:6780–6786.

McLennan, J. F. 1876. *Studies in Ancient History.* London: Bernard Quaritch.

McNeill, D. 1992. *Hand and Mind: What Gestures Reveal about Thought.* Chicago: Chicago University Press.

Mercader, J., M. Panger, and C. Boesch. 2002. Excavation of a chimpanzee stone tool site in the African rainforest. *Science* 296:1452–1455.

Mesulam, M. M. 1998. From sensation to cognition. *Brain* 121:1013–1052.

Miller, E. K. 1999. The prefrontal cortex: Complex neural properties for complex behavior. *Neuron* 22:15–17.

Milton, K. 1987. Primate diets and gut morphology: Implications for hominid evolution. In *Food and Evolution: Toward a Theory of Food Habits*, ed. M. Harris and E. B. Ross, 93–115. Philadelphia: Temple University Press.

Milton, K. 1993. Diet and primate evolution. *Scientific American* 269:86–93.

Minsky, Marvin. 2006. *The Emotion Machine: Commonsense Thinking, Artificial Intelligence, and the Future of the Human Mind*. New York: Simon & Schuster.

Mithen, S. 1996. *The Prehistory of the Mind: The Cognitive Origins of Art, Religion and Science*. London: Thames & Hudson.

Montagu, A. 1961. Neonatal and infant immaturity in man. *J. Amer. Med. Assoc.* 178:56–57.

Montagu, A. 1964. *Life Before Birth*. New York: The New American Library.

Morris, D. J. 1967. *The Naked Ape: A Zoologist's Study of the Human Animal*. London: Jonathan Cape.

Mountcastle, V. 1978. An organizing principle for cerebral function: The unit model and the distributed system. In *The Mindful Brain*, ed. G. M. Edelman and V. Mountcastle, 7–50. Cambridge, MA: MIT Press.

Neville, H. J., D. Bavelier, D. Corina, J. Rauschecker, A. Karni, A. Lalwani, A. Braun, V. Clark, P. Jezzard, and R. Turner. 1998. Cerebral organization for deaf and hearing subjects: Biological constraints and effects of experience. *Proc. Natl. Acad. Sci. USA* 95:922–929.

Nishida, T. 1990. *The Chimpanzees of the Mahale Mountains: Sexual and Life History Strategies*. Tokyo: Tokyo University Press.

Nishida, T., M. Hiraiwa-Hasegawa, T. Hasegawa, and Y. Takahata. 1985. Group extinction and female transfer in wild Chimpanzees in the Mahale National Park, Tanzania. *Z. Tierpsychol.* 67:284–301.

Odling-Smee F.J., Laland K. N. & Feldman M.W. 2003. *Niche Construction: The Neglected Process in Evolution*. Princeton: Princeton University Press.

O'Toole, K. 1999. Psychologist Zajonc urges interdisciplinary research on massacres, genocide. Stanford Online Report. May 5, 1999. http://www. stanford.edu/dept/news/report/news/may5/ massacre-55.html.

Palmer, D. 2006. *Seven Million Years: The Story of Human Evolution*. London: Phoenix.

Parker, S. T. 1990. Why big brains are so rare: Energy costs of intelligence and brain size in anthropoid primates. In *Language and Intelligence in Monkeys and Apes*, ed. S. T. Parker and K. R. Gibson, 129–156. Cambridge: Cambridge University Press.

Passingham, R. E. 1982. *The Human Primate*. Oxford: W. H. Freeman.

Pasternak, C. 2004. *Quest: The Essence of Humanity*. Chichester, UK: Wiley.

Patterson, R., and E. Linden. 1981. *The Education of Koko*. New York: Holt, Rinehart and Winston.

Penfield, W., and H. H. Jasper. 1954. *Epilepsy and the Functional Anatomy of the Human Brain*. Boston: Little, Brown.

Pianka, E. R. 1970. On r- and K-selection. *American Naturalist* 104:592–597.

Pickering, T. R., T. D. White, and N. Toth. 2000. Cutmarks on a Plio-Pleistocene hominid from Sterkfontein, South Africa. *American Journal of Physical Anthropology* 111:579–584.

Pilbeam, D. 1986. Distinguished lecture: Hominoid evolution and hominid origins. *Amer. Anthropologist*. 52:295–312.

Pinker, S. 1994. *The Language Instinct: How the Mind Creates Language*. New York: Morrow.

Pinker, S. 1997. *How the Mind Works*. New York: Norton.

Portmann, A. 1945. Die Ontogenese des Menschen als Problem der Evolutionsforschung. *Verb. Schweiz. Naturf. Ges.* 125:44–53.

Potts, R. 1996. *Humanity's Descent: The Consequences of Ecological Instability*. New York: Morrow.

Power, M. 1991. *The Egalitarians—Human and Chimpanzee*. Cambridge: Cambridge University Press.

Powers, W. T. 1973. *Behavior: The Control of Perception*. Chicago: Aldine.

Powers, W. T. 2008. *Living Control Systems III: The Fact of Control*. New Canaan, CT: Benchmark Publications.

Premack, D. 1976. *Language and Intelligence in Ape and Man*. Hillsdale, NJ: Erlbaum.

Premack, D., and A. Premack. 2003. *Original Intelligence: Unlocking the Mystery of Who We Are*. New York: McGraw-Hill.

Pretty, J. 1998. Feeding the world with sustainable farming or GMOs? *Splice* 4:4–5.

Proust, J. 1999. Indexes for action. *Revue Internationale de Philosophie, Neurosciences* 3:321–345.

Pulvermüller, F. 1999. Words in the brain's language. *Behavioral and Brain Sciences* 22:253–336.

Pulvermüller, F., M. Hare, and F. Hummel. 2000. Neurophysiological distinction of verb categories. *Cognitive Neuroscience* 11:2789–2793.

Pusey, A., J. Williams, and J. Goodall. 1997. The influence of dominance rank on the reproductive success of female chimpanzees. *Science* 277:828–831.

Reston, J. 1994. *Galileo: A Life*. New York: HarperCollins.

Richerson, P. J., and R. Boyd. 1998. The evolution of human ultra-sociality. In *Indoctrinability, Ideology and Warfare,* ed. I. Eibl-Eibesfeldt and F. Salter, 71–95. London: Berghahn Books.

Richmond, B. G., D. R. Begun, and D. S. Strait. 2001. Origin of human bipedalism: The knuckle-walking hypothesis revisited. *Yearb. Phys. Anthropol.* 44:70–105.

Ridley, M. 1993. *The Red Queen: Sex and the Evolution of Human Nature*. New York: Macmillan.

Ridley, M. 2003. *Nature Via Nurture: Genes, Experience, and What Makes Us Human*. New York: HarperCollins.

Robins, R. H. 1979. *A Short History of Linguistics*. 2nd ed. London: Longman.

Rosenberg, K. R., and W. R. Trevathan. 1996. Bipedalism and human birth: The obstetrical dilemma revisited. *Evolutionary Anthropology* 4:161–168.

Rosenbrock, H. 1990. *Machines with a Purpose*. Oxford: Oxford University Press.

Rosset, P., J. Collins, and F. M. Lappé. 2000. Lessons from the green revolution: Do we need new technology to end hunger? *Tikkun Magazine* 15:52–56.

Rozin, P., J. Haidt, and C. R. McCauley. 1993. Disgust. In *Handbook of Emotions*, ed. M. Lewis, and J. M. Haviland, 69–73. New York: Guilford.

Rumbaugh, D. M. 1977. *Language Learning by a Chimpanzee: The Lana Project*. New York: Academic Press.

Savage-Rumbaugh, E. S. 1986. *Ape Language: From Conditioned Response to Symbol*. New York: Columbia University Press.

Savage-Rumbaugh, E. S., S. G. Shanker, T. J. Taylor. 1998. *Apes, Language, and the Human Mind*. New York: Oxford University Press.

Saygin, A.P., I. Cicekli, and V. Akman. 2000. Turing test: 50 years later. *Minds and Machines* 10:463–518.

Schjelderup-Ebbe, T. 1935. Social behavior of birds. In *Handbook of Social Psychology*, ed. C. A. Murchison, 947–972. Worcester, MA: Clark University Press.

Schroeder, C. E., and Foxe, J. 2005. Multisensory contributions to low-level, "unisensory" processing. *Curr. Opin. Neurobiol.* 15:454–458.

Seabright, P. 2004. *The Company of Strangers: A Natural History of Economic Life*. Princeton: Princeton University Press.

Senut, B., M. Pickford, D. Gommery, P. Mein, K. Cheboi, and Y. Coppens. 2001. First hominid from the Miocene (Lukeino Formation, Kenya). *C. R. Acad. Sci. Paris* 332:137–144.

Shannon, C. E. 1948. A mathematical theory of communication. *Bell System Tech. J.* 27:379–423, 623–656.

Shannon, C. E., and Weaver, W. 1949. *The Mathematical Theory of Communication.* Urbana, IL: University of Illinois Press.

Shepher, J. 1971. Mate selection among second generation kibbutz adolescents and adults: Incest avoidance and negative imprinting. *Arch. Sex. Behav.* 1:293–307.

Shilts, R. 1987. *And the Band Played On: Politics, People, and the AIDS Epidemic.* New York: St. Martin's Press.

Sillito, A., and H. E. Jones. 2002. Corticothalamic interaction in the transfer of visual information. *Phil. Trans. R. Soc. Lond.* B. 357:1739–1752.

Slurink, P. 1993. Ecological dominance and the final sprint in hominid evolution. *Human Evolution* 8:265–73.

Smith, D. L. 2004. *Why We Lie: The Evolutionary Roots of Deception and the Unconscious Mind.* New York: St. Martin's Press.

Smith, D. L. 2007. *The Most Dangerous Animal: Human Nature and the Origins of War.* New York: St. Martins Press.

Sober, E., and D. S. Wilson. 1998. *Unto Others: The Evolution and Psychology of Unselfish Behavior.* Cambridge, MA: Harvard University Press.

Sounds True, Inc. 2008. *Measuring the Immeasurable: The Scientific Case for Spirituality.* Boulder, CO: Sounds True, Inc.

Sponheimer, M., and J. A. Lee-Thorp. 1999. Isotopic evidence for the diet of an early hominid, *Australopithecus africanus. Science* 283:368–370.

Staub, E. 1989. *The Roots of Evil: The Origins of Genocide and Other Group Violence.* Cambridge: Cambridge University Press.

Sterelny, K. 2003. *Thought in a Hostile world: The Evolution of Human Cognition.* Oxford: Blackwell.

Stewart, I., and J. Cohen. 1997. *Figments of Reality: The Evolution of the Curious Mind.* Cambridge: Cambridge University Press.

Sutherland, S. 1996. *The International Dictionary of Psychology.* New York: Crossroads.

Symons, D. 1979. *The Evolution of Human Sexuality.* Oxford: Oxford University Press.

Tague, R. G., and C. O. Lovejoy. 1986. The obstetric pelvis of A.L. 288-1 (Lucy). *J. Human Evolution* 15:237–255.

Tanner, N. M. 1981. *On Becoming Human*. Cambridge: Cambridge University Press.

Tattersall, I. 1995. *The Last Neanderthal: The Origin, Success, and Mysterious Extinction of Our Closest Human Relative*. New York: Macmillan.

Tattersall, I. 1998. *Becoming Human: Evolution and Human Uniqueness*. New York: Harcourt Brace.

Tolman, E. C. 1932. *Purposive Behavior in Animals and Men*. New York: Century.

Tooby, J., and L. Cosmides. 1988. The evolution of war and its cognitive foundations. *Proc. Inst. Evolutionary Studies* 88:1–15.

Trevathan, W. R. 1987. *Human Birth: An Evolutionary Perspective*. New York: Aldine de Gruyter.

Trinkaus, E. 1983. *The Shanidar Neandertals*. New York: Academic Press.

Trivers, R. L. 1971. The evolution of reciprocal altruism. *Quarterly Review of Biology* 46:35–57.

Trivers, R. L. 1972. Parental investment and sexual selection. In *Sexual Selection and the Descent of Man: 1871–1971*, ed. B. G. Campbell, 136–179. Chicago: Aldine.

Turing, A. M. 1950. Computing machinery and intelligence. *Mind* 59:433–460.

Tylor, E. B. 1871. *Primitive Culture: Researches into the Development of Mythology, Philosophy, Religion, Language, Art and Customs*. London: John Murray.

van der Dennen, J. M. G. 1995. *The Origin of War: The Evolution of a Male-Coalitional Reproductive Strategy*. Groningen: Origin Press.

van Schaik, C. P., and P. M. Kappeler. 1997. Infanticide risk and the evolution of male-female association in primates. *Proc. Royal Soc. Lond. B* 264:1687–1694.

van Schaik, C. P., and R. I. M. Dunbar. 1990. The evolution of monogamy in large primates: A new hypothesis and some crucial tests. *Behaviour* 115:51–56.

Wade, N. 2006. *Before the Dawn: Recovering the Lost History of Our Ancestors*. New York: Penguin.

Wall-Scheffler, C. M., K. Geiger, and K. L. Steudel-Numbers. 2007. Infant carrying: The role of increased locomotory costs in early tool development. *Amer. J. Phys. Anthropol.* 133:841–846.

Walker, A. and P. Shipman. 1996. *The Wisdom of the Bones*. Knopf: New York.

Wallace, A. R. 1869. *The Malay Archipelago: The Land of the Oran-utan and the Bird of Paradise, a Narrative Travel with Studies of Man and Nature*. London: Macmillan.

Waller, J. 2002. *Becoming Evil: How Ordinary People Commit Genocide and Mass Killing*. New York: Oxford University Press.

Warren, J. M., and K. Akert, eds. 1964. *The Frontal Granular Cortex and Behavior*. New York: McGraw-Hill.

Waters, J. D. 2006. *Helpless as a Baby*. Bloominton, IN: AuthorHouse.

Wendorf, F. 1968. Site 117: A Nubian final paleolithic graveyard near Jebel Sahaba, Sudan. In *The Prehistory of Nubia*, ed. F. Wendorf, 954–995. Dallas: Southern Methodist University Press.

Wexler, M., S. M. Kosslyn, and A. Berthoz. 1998. Motor processes in mental rotation. *Cognition* 68:77–94.

White, T. D., G. Suwa, and B. Asfaw. 1994. Australopithecus ramidus, a new species of hominid from Aramis, Ethiopia. *Nature* 371:306–312.

Whiten, A. W., and R. W. Byrne, eds. 1988. *Machiavellian Intelligence: Social Expertise and the Evolution of Intellect in Monkeys, Apes, and Humans*. New York: Oxford University Press.

Whiten, A. W., and R. W. Byrne, eds. 1997. *Machiavellian Intelligence II: Extensions and Evaluations*. Cambridge: Cambridge University Press.

Whiten, A., and C. Boesch. 2001. The culture of chimpanzees. *Scientific American* 284:61–67.

Whiten, A., J. Goodall, W. C. McGrew, T. Nishida, V. Reynolds, Y. Sugiyama, C. E. G. Tutin, R. W. Wrangham, and C. Boesch. 1999. Cultures in chimpanzees. *Nature* 399:682–685.

Williams, G. C. 1966. *Adaptation and Natural Selection: A Critique of Some Current Evolutionary Thought.* Princeton: Princeton University Press.

Williams, G. C. 1975. *Sex and Evolution.* Princeton: Princeton University Press.

Wilson, D. S. 1975. A general theory of group selection. *Proc. Natl. Acad. Sci. USA* 72:143–146.

Wilson, D. S., and E. Sober. 1994. Reintroducing group selection to the human behavioral sciences. *Behav. Brain Sci.* 17:585–654.

Wilson, E. O. 1975. *Sociobiology: The New Synthesis.* Cambridge, MA: Harvard University Press.

Wolf, A. P. 1968. Adopt a daughter-in-law, marry a sister: A Chinese solution to the problem of the incest taboo. *American Anthropologist* 70:864–874.

Wolf, A. P. 1995. *Sexual Attraction and Childhood Association: A Chinese Brief for Edward Westermarck.* Stanford, CA: Stanford University Press.

Wrangham, R. W. 1987. Evolution of social structure. In *Primate Societies,* ed. B. B. Smuts, D. L. Cheney, R. Seyfarth, R. W. Wrangham, and T. T. Struhsaker, 282–296. Chicago: University of Chicago Press.

Wrangham, R. W., and D. Peterson. 1996. *Demonic Males: Apes and the Origins of Human Violence.* Boston: Houghton Mifflin.

Wrangham, R. W., W. C. McGrew, F. B. M. de Waal, and P. G. Heltne, eds. 1994. *Chimpanzee Cultures.* Cambridge, MA: Harvard University Press.

Wynne-Edwards, V. C. 1962. *Animal Dispersion in Relation to Social Behavior.* Edinburgh: Oliver and Boyd.

Zajonc, R. B. 2002. The zoomorphism of human collective violence. In *Understanding Genocide: The Social Psychology of the Holocaust,* eds. L. S. Newman and R. Erber, 222–240. New York: Oxford University Press.

Zeki, S. 1999. *Inner Vision: An Exploration of Art and the Brain.* New York: Oxford University Press.

INDEX

A

action
  in aggression 121, 140, 141, 146,
    161, 181, 222
  in altruism 160
  in cognition 28, 30, 31, 32, 33, 34,
    35, 39, 41, 46, 47, 50, 51, 89,
    90, 94
  in control theory 7, 10, 11, 12, 21,
    42, 44, 69, 92, 110, 111, 113,
    161, 205
  in cultural signaling 183
  in syntactic language 81, 83, 87, 89,
    93, 97, 224
  in technology 112, 113, 119
  in the arts 190, 194
  in uncontrollable situations 203, 205,
    207, 209, 213
  neural systems 32, 37, 38, 40, 44, 55,
    85, 86, 111
  selective inhibition in altruism 160,
    162
  shared in human organizations 74
  species comparisons 17
afterlife 209, 211, 214, 227
aggression 25, 121, 146, 160, 163,
    164, 165, 184, 214, 215, 222,
    226, 228
  and technology 221
  co-evolutionary 148, 149, 163
  human patterns 122, 124, 138, 139,
    140, 141, 142, 143, 146, 158,
    184, 221, 226
  in control theory 121, 140, 141, 161,
    163, 222
  in dominance hierarchy 135
  inlawship/friendship 132, 154
  kinship 130, 131, 133, 152

  reproductive 127, 151
  social dimensions 125, 135, 149
  species comparisons 122, 137, 138
  threat of extinction 122, 124, 144,
    148, 222, 223
  xenophobic 135, 136, 137, 165, 174,
    224
altruism 25, 123, 144, 145, 146, 157,
    164, 165, 184, 222, 228
  co-evolutionary 147, 149, 163
  contrast with selfishness 146
  cultural 146, 159, 160
  friendship 154, 156
  group selected 157, 158, 159
  in control theory 160, 161, 162, 163
  inlaw 153, 154
  kinship 151, 152, 155
  protection against extinction 148,
    149, 163, 164, 214, 223
  reciprocal 154, 155, 156
  reproductive 150, 151, 164
  social dimensions 149, 165, 184, 223
  sociobiological models 146, 222, 223
  warrior 159
American Sign Language 76, 81
aphasias 85
arts 25, 166, 184, 187, 189, 195, 197,
    198
  appreciation of art 191, 192, 195,
    198
  cave paintings 187, 194
  emergence of artists 187, 189, 190,
    198
  in control theory 192, 193, 195, 196,
    197, 198
  limitations of cultural memetics 187,
    188
atrocity 121, 138, 139, 140, 144

among chimpanzees 137
as control behavior 142, 143, 165, 221, 222
restraint by cultural signaling 226
restraint by religion 228
Australopithecus 100, 101, 112, 116, 218, 219

B

Bickerton, Derek 78, 79, 80, 81, 83
bipedal locomotion 25, 99, 100, 101, 167, 168, 219
co-evolutionary 102, 104, 105, 108, 120
hominid trinity 106, 107, 108, 219
infant care 101, 105, 106, 219
obstetrical complications 101
species comparisons 100, 102, 108, 219
transportation economy 109, 120, 155
blank slate 77, 168
bonobo 3, 5, 6, 99, 124, 166, 217
brain
aphasias 85
cell assemblies 86
development 49, 50, 54, 103, 104, 105, 106, 107, 108, 168, 219, 224
expansion 25, 52, 54, 67, 84, 100, 101, 104, 105, 107, 108, 116, 118, 119, 120, 164, 172, 180
in action 37
in art 191, 193, 197
in cognition 33, 40, 52, 56, 114, 142, 189, 224
in control theory 41, 98, 111, 160, 182
in emotion 68
in meme theory 171, 188
in perception 36, 37
in prediction 46, 48, 215
in self-awareness 56, 58, 63, 64, 66, 67

in spirituality 202, 207, 215
in syntactic language 74, 78, 84, 85, 86, 89, 93, 97, 224
metabolism 54, 103, 106, 115, 117, 120
obstetrical complications 104
recurrent loops 64, 65, 66
scanning techniques 59, 84, 85, 86
species comparisons 6, 26, 52, 53, 54, 67, 98, 100, 103, 108, 112
structure 19, 22, 23, 26, 29, 35, 39, 40, 52, 53, 58, 60, 64, 75, 76, 77, 84, 95, 99, 104, 116, 119, 120, 184, 212, 219

C

cave paintings 4, 187, 194, 195
cell assemblies 65, 86, 87, 93
cerebral cortex 38, 39, 52, 53, 54, 55, 66, 70, 71, 85, 189
childbirth 127
chimpanzee 3, 25, 151, 216, 217, 220
action 32
aggression 122, 136, 137, 138, 139, 143, 222
altrusim 155
anthropomorphization 13
brain 53, 54, 100, 103
cognition 27, 32
common ancestor 6, 216, 217
culture 169, 184, 225
death 203
hunting 118
language 80, 82, 83, 218
locomotion 99, 109
meat eating 116
parenting behavior 128
perception 32
post-natal development 104, 105
self-awareness 59, 64, 71
sexual/reproductive behavior 127, 134, 137
social system 96, 124, 127, 134, 155, 165, 176, 218

survival strategy 5, 6, 218
tool-use 109, 114
xenophobia 165, 218
circularity 9, 40, 60, 65, 67, 111, 112
classical sandwich 33, 34, 35
climate change 2, 4
  oscillation 13, 15, 16, 22, 23
clinging behavior 101, 102, 105
cognition 25, 27, 28, 31, 33, 55, 75
  action 30
  adaptive value 27, 28, 45, 224
  as information processing 29, 31, 32,
    34, 39, 41, 52, 56, 110, 155
  classical sandwich 33, 41
  control models 41, 45, 46, 49, 51,
    89, 92, 93, 94, 95, 97, 141, 192,
    197, 207
  emotional perspectives 68
  experiential perspectives 56, 65, 67
  hierarchical models 35, 36, 37, 38,
    39, 41, 45
  in aggression 142, 146
  in memetic theory 171
  in reciprocal altruism 156
  in spirituality 205, 206, 207, 209,
    215, 227
  in technology 111, 112, 118, 119,
    120
  interactive models 35, 39, 40, 55,
    89, 97
  in the arts 192, 195, 197
  linkages 31, 32, 33, 34, 35, 38, 41,
    44
  neural systems 29, 38, 54
  neurophilosophical models 34, 35,
    38, 41, 48
  perception 29
  predictive functions 46, 48, 49, 203
  research approaches 28, 29, 33
  species comparisons 27, 32, 55, 155
  syntactic language 74, 75, 77, 80, 87,
    89, 90, 93, 94, 95, 96, 97
  tabula rasa models 77
cognitive objectification 142
cognitive science 33, 68, 76

consciousness. See self-awareness
consummatory behavior 111, 112,
  126, 127
control
  as human strategy xiii, xiv, 3, 5, 82,
    104, 120, 121, 160, 183, 216,
    217, 221, 231
  comparator function 10, 11, 42, 44,
    46, 111
  cooperative organizations 95, 97
  efference copy effects 48
  electronic circuits 17
  equivalence principle 66, 67
  evolutionary path 19, 20, 21, 23, 25,
    219, 221, 231
  feedback 64
  in aggression 121, 122, 140, 141,
    142, 143, 144, 146, 148, 159,
    164, 222, 223
  in altruism 149, 160, 161, 163, 180,
    223
  in biological circuits 19
  in cognition 40, 41, 45, 49, 52, 55,
    56, 90, 224
  in cultural signaling 183, 185, 226
  in deception 184
  in emotions 68, 69
  in religion 212, 213, 214, 227
  in science 211, 228, 229, 230
  in self-awareness 56, 65, 70
  in spirituality 200, 206, 207, 209,
    210, 212, 215, 227
  in syntactic language 73, 74, 75, 84,
    89, 95, 97, 224
  in technology 111, 119
  in the arts 191, 192, 193, 197
  in uncontrollable situations 201, 202,
    207
  multi-level arrays 42, 44
  neural systems 66, 98, 184
  operational definition 6, 7, 8, 9, 10,
    11, 12, 17, 160
  predictive functions 46, 48, 49, 50,
    51
  ratcheting principle 113

replicable processes 111
species comparisons 6, 55, 217, 218
stabilization 13, 17
conversation 72, 96, 97, 218
cooperation
  and aggression 140
  and xenophobia 174
  in altruism 147, 149, 154, 156
  in control theory 74
  promoted by cultural signaling 179,
    183, 186
  social structure 123, 124, 125, 127,
    130, 131, 132, 133, 137, 184,
    224
  species comparisons 96, 218, 220,
    224
  survival strategy 118, 123
corpse 204, 205
correlation problem 49
Creole languages 79
Crick, Francis 66
cultural signaling 144, 173, 174, 175,
    179, 185, 226
  added social dimension 179
  between SKIFs 178, 179, 183, 186,
    225
  in deception 184, 185
  in the arts 188
  trivialities 225
culture 25, 123, 172, 184, 222
  contrast to instinct 167
  definition 166, 167, 169, 188, 224
  evolutionary path 171, 172, 173, 226
  exchanges between groups 173, 174
  in child development 168, 169
  in control theory 180
  increasing aggression 174
  in death 204, 205
  in religion 213, 214, 226, 228
  meme theory 170, 171
  nature-nurture 167
  origins of deception 184
  reducing xenophobia 166, 178, 214,
    226

signaling between groups 176, 178,
    183, 225, 226
signaling within groups 175
social dimension 224
source of altruism 159, 160
source of selfishness 160
source of ultra-sociality 166
species comparisons 114, 169, 184,
    205, 224, 225
the arts as a special case 187
trivialities 172, 225

D

Darwinian Theory 14, 19, 133, 157,
    216
Dawkins, Richard 129, 170, 188
death 215, 227
  as a control problem 202, 203, 204,
    205, 207, 209, 214
  funeral practices 205
  human premonition 202
  impetus for spirituality 206, 209, 212
  in chimpanzees 203
  premonition 203
  test of religion 213, 214
deception 184, 186
dehumanization 142
diet 116, 120
  and brain expansion 115
  meat preference 116, 117
digestive system 116
dissonance
  in aggression 140, 141, 143
  in altruism 162
  in control theory 10, 11, 12, 21, 42,
    44
  in cultural signaling 181, 183
  in death 203, 204, 205
  in deception 184
  in emotions 69
  in process improvements 113
  in spirituality 207, 209, 214, 215,
    227
  in syntactic language 90, 92, 94

in the arts 192, 193, 194, 195
lateral connectivity 44
disturbances
    in control theory 11, 12
    in process improvements 113
dominance-subordinance hierarchy 96,
        132, 134, 135, 173
Dunbar, Robin 95, 96, 165, 174

E

Edelman, Gerald 65
effects
    in control theory 10, 11, 44, 47
efference copy 48, 49
egalitarian 133, 134, 135, 220
emotions 67, 68
    adaptive value 68
    as varieties of self-awareness 67
    as ways of thinking 68
    correlates in dissonance patterns 69
    in the arts 198
epilepsy 66
equivalence principle 66, 67, 68
expectation
    evolutionary path 17, 19, 20, 21, 22,
        23, 98, 216, 218, 219
    forming the transportation economy
        110, 219
    in aggression 141, 143
    in altruism 160
    in child development 50
    in cognition 90
    in control theory 7, 8, 10, 11, 12, 17,
        19, 44, 74, 75, 92, 98, 219
    in cultural development 173
    in cultural signaling 180, 181, 182,
        183
    in death 202, 204, 205
    in emotions 69
    in multi-level control arrays 44, 45
    in process improvements 111, 113
    in self-awareness 70
    in spirituality 207
    in syntactic language 94, 97

in technology 112, 119
in the arts 192, 193, 195, 196, 197,
        198
in using tools 111
match/mismatch with perception 11,
        12
neural systems 8, 9, 10, 12, 19, 25,
        182
of immortality 203, 209
programmed sequences 12
shared in human organizations 73,
        74, 75, 90, 95, 218, 224
species comparisons 22
extinction 25
    and climate change 14, 15
    and climate oscillation 16
    and inflexible behavior 23
    infant mortality 120
    in group selection theory 158
    in memetic theory 188
    of altruists 148, 158, 159
    precursors 122, 130, 131, 133, 148,
        223
    prevention by altruism 148, 161, 163

F

feedback
    biological circuits 19
    correlation problem 49
    efference copy mechanisms 48
    electronic circuits 17
    imagination connection 46
    in adult learning 51
    in biological circuits 66
    in cell assemblies 65, 66
    in childhood development 50
    in control theory 11, 41, 64
    in predictive functions 46, 48, 49
    in self-awareness loops 62, 64, 70
    in spiritual hypothesis building 207,
        209
    in syntactic language 92
    lateral connectivity 44
    sensory-motor reflexes 38

friendship
  in death 205
  in reciprocal altruism 154, 155
  social structure 124, 131, 132, 136,
    165, 175
funeral practices 204

G

genetic diversity
  avoidance of incest 130
  evolution of altruism 157, 223
  evolution of bipedalism 107
  evolution of control systems 19, 216
  precursor to extinction 122, 148,
    223, 229
  requirement for evolution 14, 15,
    129
  social implications 130, 133, 134,
    137, 152, 173, 224
genocide 5, 121, 139, 144
  among chimpanzees 143
  as control behavior 140, 142, 143,
    165, 222
  precursor to extinction 148, 164, 223
  restraint by cultural signaling 186,
    226
  restraint by religion 228
gestures
  in cultural signaling 225
  in language 75, 80, 82, 93
Gombe 114, 136, 137, 138, 139, 143
grammar 72, 75, 76
  childhood development 78, 79
  evolution 77, 79, 82
  in cognition 89
  in protolanguage 78
  universal 76, 77, 79
grooming 95
group selection 157, 158, 159

H

Hebb, Donald 65, 86, 87
hierarchical organizations 35, 36
  action 37, 38
    complications 36, 39
    in cognition 35, 36
    in control theory 41, 42, 44, 45
    in predictive functions 48
    lateral connectivity 40
    perceptual 36, 37
    reciprocal connectivity 39
Hofstadter, Douglas 60
Hominid Trinity 107, 219
Homo aestheticus 197, 198
Homo auguris 27, 28, 32, 45, 49, 55,
    112
Homo bellicosus 121, 122, 143, 144,
    146, 148, 162, 163, 186, 215
Homo beneficus 145, 160, 163
Homo conlocutus 72, 75, 97
Homo dominus xiv, 1, 5, 6, 13, 25, 26,
    186, 191, 201, 216, 217, 223,
    226, 227, 228, 230, 231
Homo erectus 1, 71, 139, 220, 229
Homo habilis 98, 99, 112, 114, 119,
    120
Homo humanitas 165, 185
Homo ipsianimus 56, 60, 64, 70, 71
Homo mortalis 199, 210, 212, 215
Homo sapiens xiv, 5, 216, 217, 219,
    223, 226, 228, 230, 231
humanness xiii, 1, 4, 5, 12, 13, 17, 23,
    29, 45, 97, 107, 108, 218, 219,
    220, 231
human signature xiv, 25
human strategy 3, 5, 26, 95, 110, 113,
    121, 164, 215, 218
hunting 116, 118, 143, 156, 157, 167,
    168
Hurley, Susan 34, 35, 41, 48

I

i-culture 170
imagination 7, 41, 46, 207
imitation 170, 171, 172, 188, 203, 213
incest 130, 131, 134
inclusive fitness
  in altruism 151, 153

social implications 129, 130, 131,
133, 224
infant helplessness
brain expansion 103
evolutionary path 105, 106
impetus for bipedal locomotion 107,
108, 219
precursor to extinction 103
social implications 150
information processing
in cognition 28, 29, 30, 31, 32
in control theory 41, 44, 56
in self-awareness 65
in syntactic language 84
in the arts 197
neural systems 52
inlawship
aggression potential 132
social structure 124, 131, 132, 136,
165, 175
instinct
consummatory behavior 127
in aggression 148, 226
in altruism 223
in control theory 9, 10, 44
in cultural development 169
in death 204
infant care 101, 105, 109, 128
in spirituality 203, 204, 206, 207,
215
in syntactic language 77, 79, 82
in the arts 191
mating preferences 130
nature-nurture 167, 168, 169
intrinsic value 110, 155

K

kin selection 129, 131, 151, 152
kinship
aggression potential 131, 132, 133,
136, 137, 139, 151
in altruism 152, 153, 154, 165
incest avoidance 130
in death 205

selfish gene theory 129
social structure 124, 127, 128, 129,
130, 132, 175, 176
K-selection 128, 129

L

language 4, 73, 75, 76, 77, 168, 169
adaptive value 95, 96, 97
and conversation 72
and grooming 95
aphasias 85
childhood development 79
evolution 75, 76, 77, 78
gestural forms 13, 80
in cognition 89, 95, 97
neural systems 86, 89
scanning techniques 85
lateral connectivity 39, 40
learning
based on imitation 171
in control theory 9
in cultural development 166, 167,
168, 171, 224, 225
in cultural signaling 225
predictive correlations 49, 51
lexicons 97
linguistics 76
Linnaeus xiv, 5, 26
loop structures
cell assemblies 65, 87
external closure 9, 10, 46
foundation of self-awareness 60, 61,
62, 63, 64, 65, 70
in control theory 9, 10, 11, 41, 69
in emotions 67, 69
in hierarchical systems 36, 40
in predictive functions 47
in spirituality 207
in syntactic language 92, 95
internal closure 11, 42, 46
neural systems 38, 66
Lucy 100, 101

M

magic 200, 201
m-culture 170
meat
  and brain expansion 115
  and technology 117
  primate diet 115, 116
  scavenging 117
memes 169
  basis of cultural transmission 170,
    172, 188
  childhood development 173
  in the arts 188
  in xenophobic environments 173
  modification during transmission
    171, 180
  trivialities 172
  viral 184, 185
memetic theory 170, 171, 178, 214
memory
  predictive correlations 49, 51
migration 2, 139
militarism 159
Miocene 1, 2, 13, 14, 15, 16, 19, 23,
  28, 99, 102, 223
mirror test 59
modular theory 50, 51
mutation
  basis of adaptive evolution 14, 19
  brain expansion 104
  in control systems 21, 22, 23
  in slowly reproducing species 15, 22
  response to ecological oscillation 16,
    22, 23

N

nationalism 159, 179
natural selection
  and selfish genes 170
  at the group level 157
  for adaptive action 28
  for aggression 141, 144
  for altruism 151
  for cognition 29, 52

for control stability 16
for emotions 68
for enhanced expectation 19
foundation of evolution 3, 130, 216,
  217, 231
replacement by cultural evolution
  229
nature-nurture 167, 169
nervous system 7, 35, 52, 63, 100
nouns 78
  combinatorial power 83
  in cell assemblies 87, 93
  in cognition 97
  in control theory 95
  in syntactic language 78, 81, 82, 89,
    93
  neural systems 85, 86

P

pair-bonding
  adaptive value 127, 129
  in altruism 151, 153, 164
  infant helplessness 151
  potential for aggression 151
  social implications 134, 140, 152
  social structure 126, 132, 136, 176
  species comparisons 126, 127, 129,
    134
parenting 127, 128, 154
  impact of infant helplessness 129,
    150
  species comparisons 129
perception
  alteration by external feedback 10,
    44, 50, 161
  alteration by internal feedback 11,
    46, 49, 51, 92, 207
  augmentation in the arts 189, 190,
    191, 192, 193, 195
  childhood development 50
  comparator function 9, 10
  efference copy mechanisms 48
  hierarchical models 36, 37
  hypothetical 206

imagined 7
in aggression 121, 141, 161
in altruism 160, 162
in cell assemblies 87, 93
in cognition 28, 31, 32, 33, 34, 35,
    39, 40, 45, 46, 50, 51, 89, 90,
    94
in control theory 7, 8, 9, 10, 11, 12,
    20, 42, 44, 74, 75, 92, 219
in cultural signaling 181, 182, 183
in multi-level control arrays 44
in predictive functions 46, 47
in process improvements 111, 113
in self-awareness 70
in spirituality 205, 207
in syntactic language 83, 87, 89, 90,
    93, 97, 224
in technology 112, 119
in the arts 192, 193, 194, 195, 196,
    197, 198
in the transportation economy 110
in using tools 111
lateral connectivity 40
match/mismatch with expectation
    11, 12, 21, 69, 94
neural systems 35, 36, 38, 55, 85,
    86, 93
of death 203, 204
predictive correlation problem 49, 50
reciprocal connectivity 39
shared in human organizations 74,
    224
species comparisons 17, 22, 23, 32
perception-expectation match/mismatch
    10, 11, 12, 21, 22, 69, 160, 161,
    181, 182, 192, 193, 195, 196,
    203, 207, 209
phonetics 75
pidgin languages 78, 79
post-fertilization activity 126, 127
Powers, William 41, 46, 207
precocial 102, 103
prediction 25, 41, 46, 49
  correlation problem 51
  efference copy mechanism 48

in technology 112
internal feedback mechanisms 46, 47,
    48, 208
prefrontal cortex 38, 53
prenuptial signaling 126
protolanguage 78, 80, 81, 83
pseudospeciation 142, 143

R

racism 179
ratcheting principle 113, 114, 213,
    224, 225
reciprocal connectivity 36, 39
recurrent structures
  cell assemblies 65, 66
  in control theory 9
  in emotions 67
  in self-awareness 61, 62, 63, 64, 66,
      67, 70
  in syntactic language 95
  neural systems 66
religion 25, 200, 214, 228
  and spirituality 207, 213, 215, 226,
      227
  rivalry with science 210, 211, 215,
      227, 228
  viral meme theory 214
reproductive fitness
  brain expansion 101
  dominance-subordinance 133, 135
  ecological impacts 14, 16, 17
  foundation of evolution 14, 19, 125,
      216
  impact of pair-bonding 134, 150,
      151
  impact of science 229
  in aggression 122, 123, 127, 144,
      148, 164, 223
  in altruism 146, 148, 150, 152, 157,
      158, 159
  in bipedalism 106, 107
  incest avoidance 130, 134, 137
  in cognition 28, 31
  in control theory 23

in cultural development 171, 174
in emotion 68, 69
in group selection theory 157
in prediction 46, 49
in syntactic language 75
kin selection 129, 151
selfish gene theory 130, 170
sexual selection 125, 126
social implications 123, 126, 127,
    128, 129, 130, 132, 134, 150
reproductive rates 15, 16, 103
r-selection 128

S

scavenging 83, 117, 118
science 25, 58, 228, 229
  and spirituality 215, 227
  rivalry with religion 210, 211, 215,
    227, 228
secondary altriciality
  brain expansion 102, 104
  delayed postnatal development 103
  evolutionary path 104
  hominid trinity 106, 219
  social implications 128, 129
  threat of extinction 103, 105
self-awareness 25, 29, 56, 57, 58, 64,
    66, 70, 74
  emotional aspects 67, 68
  equivalence principle 66, 67
  evolutionary timeline 71
  in control theory 13, 65
  language connections 95
  mirror test 59
  philosophical issues 57, 58, 62
  structures 60, 62, 63, 64, 66, 70
  Turing test 58
selfish genes 130, 131, 146
selfishness 160
  adaptive value 148, 150, 152, 158
  basis of aggression 146
  co-evolutionary 147, 148
  contrast to altruism 145, 146, 159
  in deception 184

in reciprocal altruism 154, 156
  parasitism on altruism 148, 158, 159,
    160
  restraint by religion 214, 228
  sociobiological models 146, 147, 148
  threat of extinction 148, 149, 157,
    158, 215
semantics 75, 76, 80, 83, 95, 97
sexual selection 102, 125, 127, 128,
    133
sign language 80
SKIF 132, 186, 224, 225
  altruism within group 157, 162, 165
  benchmark size 165, 174
  cultural development 173, 174
  cultural signaling 175
  cultural signaling between groups
    176, 178, 179, 182, 188
  cultural signaling within groups 175,
    182, 225
  fission-fusion events 136, 166, 186,
    214
  genetic diversity 137
  reproductive exchanges 137
  social structure 132, 135, 136, 176,
    179
  sources of aggression 139, 140
  xenophobic posture 135, 136, 138,
    165, 174, 176, 179
social organizations 73
  adaptive value 75, 95, 96
social systems 123, 124, 125, 130, 131,
    132, 154, 155, 161, 186, 205
sociobiology 123, 124, 146, 157, 159
speciation 126
spirituality 25, 200, 206, 210, 212,
    213, 214, 215, 226, 227, 228
Sterelny, Kim 27
superstitions 199, 200, 201
  and uncertainty 201, 202
  in religious behavior 212, 227
  in uncontrollable situations 202, 209,
    227
  surrounding death 203

syntactic language 25, 74, 75, 77, 79,
    80, 89
syntax 4, 76, 97
  adaptive value 72, 81
  childhood development 79
  in cognition 97
  in control systems 95
  in control theory 93
  in protolanguage 78
  neural mechanisms 86
  species comparisons 80
  universal 77

T

tabula rasa 77, 168
technology 25, 111, 120, 229, 230.
    See Tool-Use
  adaptive value 120
  and aggression 221
  and bipedal locomotion 120
  circular logic 112
  contrast with tool-use 111, 119
  cultural development 174
  evolutionary path 112, 113, 115, 120
  meat procurement 117, 118
  process improvement 110, 113, 114,
    118, 179, 225
  species comparisons 114
teeth 116
thinking 27, 28, 29, 56, 65, 68, 70, 90
tool-use 101, 111, 112. *See* Technol-
    ogy
  adaptive value 117
  contrast with technology 111, 119
  evolutionary path 112, 219
  in chimpanzees 114
  in cultural development 169
transformational-generative linguistics
    76
transportation economy 109, 110, 115,
    117
Turing test 58, 59

U

ultra-sociality 166, 176, 212
uncertainty 200, 201, 202, 206, 209

V

value economy 120, 156
verbs 78, 81
  combinatorial power 83
  in cell assemblies 87, 93
  in cognition 97
  in control theory 95
  in syntactic language 78, 81, 89, 93
  neural systems 85, 86
viral memes 185, 214

W

warrior altruism 159

X

xenophobia 135, 166, 186, 226
  and group selection theory 157
  childhood development 173
  cultural isolation 165, 173, 174, 225
  cultural signaling between groups
    176, 178, 179, 183, 226
  cultural signaling within groups 175,
    225
  emergence of warrior altruism 159
  impact on genetic diversity 137
  in aggression 165, 179
  in chimpanzees 136, 137
  limitations of altruism 149
  social structure 124, 135, 179, 224
  threat of extinction 174, 226

Z

Zajonc, Robert 142, 143

www.ingramcontent.com/pod-product-compliance
Lightning Source LLC
Chambersburg PA
CBHW020734180526
45163CB00001B/232